N O N L I N E A R

A G u i d e t o D i g i t a l F i l m a n d V i d e o E d i t i n g

She came up to me at the party. "I hear you're an editor. Neat. Are you a film editor or a videotape editor?"

"Film ed—" I said, but I cut myself short. Hmmmm. I stood in thought. Maybe a minute passed. The inquisitor wondered what was so thought provoking about her question. I still wasn't answering, so she ignored me and began to wax philosophically. "Editing is so amazing. You cut up all that film. Hold the little frames up to the light. You must be so patient." She saw no response from me. "What do you call those noisy projector things?"

Finally I spoke. "I don't actually edit film. . . anymore. . ."

She had a blank expression. Wrinkled up her nose. She looked like she was going to say "Oh," and walk away, but she didn't. "You mean you type on one of those colored keyboards, like in 'Broadcast News'? I've seen that before. . ."

"No. No. I do edit film projects, I mean, they start on film, like a movie, and they finish on film, like a movie, but. . . I edit electronically." I could see that I was losing her. "I do use computers to edit, like a videotape editor, but it's editing video of film — like it was film. In a film style. Yes, I'm a *digital film* editor!" She seemed to be growing excited as I grew excited in my self-discovery. I went with it. "It's not videotape editing, as you know it. It *is* film editing, but different. . . I'm editing *films*, but I'm not editing *film*, if you know what I mean." She walked away and I didn't care. I had finally uncovered exactly what it was I did.

Digital Film. Nonlinear Video. This is what this book is about.

A Guide to
Digital Film and
Video Editing

Third Edition

by
Michael **Rubin**

A BOOK FROM PLAYGROUND PRODUCTIONS, SANTA CRUZ, CALIFORNIA

TRIAD PUBLISHING COMPANY • GAINESVILLE, FLORIDA

WITHDRAWN

Printed in the United States of America
First Edition November 1991
Second Edition June 1992
Third Edition October 1995
10 9 8 7 6 5 4 3 2 1

Library of Congress Cataloging-in-Publication Data
Rubin, Michael, 1963 -
   Nonlinear : a guide to digital film and video editing / by
Michael Rubin. — 3rd ed.
      p.   cm.
   "A book from Playground Productions, Santa Cruz, California."
   Includes index.
   ISBN 0-937404-84-5
   1. Motion pictures -- Editing.   2. Video tapes -- Editing.   I. Title.
TR899. R83   1995
778.5' 235 -- dc20   95-40791

Every effort has been made to provide accurate and up-to-date information about the edit systems and equipment described in this book, utilizing information supplied by the manufacturers. The opinions given, however, are solely those of the author.

Screens and interfaces of editing systems © by their manufacturers; the names of the editing systems and equipment are registered trademarks of their manufacturers, as follows:

Published and distributed by
Triad Publishing Company
PO Drawer #13355
Gainesville, FL 32604

This book was produced entirely on a Macintosh *Quadra 950*, 32MB, 3GB Seagate *Elite*, with a Nikon *Coolscan* & HP *IIc*, using Aldus *Pagemaker 5.0* (before Adobe), and illustrated with Aldus *Freehand 4.0*, Macromedia *Freehand 5.0*, and Adobe *Photoshop 3.0*

A C K N O W L E D G M E N T S

The late spring and early summer of '91 saw this guide evolve from addenda to my *CMX 6000 Manual Supplement* to a generalized overview for all systems. Now more than 4 years from its first release, it has evolved again; through all the years, I have received help from many people.

For their technical and historical expertise, I wish to thank Tom Beams, Stan Becker, William Butler, Jerry Canciellari, Josiah Carberry, Shawn Carnahan, Gabriella Cristiani, Dick Darling, Andy Delle, Robin Dietrich, Robert Doris, Herb Dow, Adrian Ettlinger, Leigh Greenberg, Patrick Gregston, Ralph Guggenheim, Seth Haberman, Augie Hess, Barbara Koalkin, Don Kravits, Robert Lay, Harry Marks, Gary Migdal, Harry Mott, Maureen O'Connell, Tom Ohanian, Tony Schmitz, Arthur Schneider, Steve Schwartz, Tom Scott, Larry Sherwood, Martha Swetzoff, Bob Turner, Randy Ubillos, Nancy Umberger, Susan Walker, Sondra Watanabe, John V. Weaver, Kenneth Yas, and especially — a profound and special appreciation — to Ron Diamond, Dean Godshall, Scott Johnston(!), and Derek McCants.

I would also like to take a moment to remember Mark Yobs, a warm friend to many in the LA post-production community; and Robert Duffy, whose contributions to the evolution of editing systems and interfaces were remarkable; both will be missed.

Above all, I want to thank my mother, Lorna Rubin, for sharing her publishing expertise, and both my parents for their unqualified support of all my unusual projects. I would also like to extend my very warm gratitude to Mary Sauer and Steve Arnold, who brought me into this business in the first place.

Finally, for all kinds of support (as innocent bystanders, all), I want to thank my family — Sara, Gabrielle, Jacques, Joshua, Danny, Louise, Maida, Asa, Jerry, Judie, Jeff, Magnum — my friends Paul, GKP, Dave, Virginia, Sue, Holly, Bruce, Tracey, Amanda, Grant, Thomas, Kathleen, Fruitman, Paul, John, Julie, Terry, Teddy, Mark, the Petroglyph gang, and most importantly to my love and wife Jennifer, who constantly reminds me that at the end of the day, all this technology isn't really worth squat.

THE CONTRIBUTORS' STORY

In the fall of 1985 I met Ron Diamond. I had been charged with giving up-close demonstrations of nonlinear editing to interested parties in the Bay Area — in particular, showing editors, directors and producers the EditDroid and SoundDroid products. Although demos technically fell under "sales" activities, these demos were more of a circus event. My job was really to get people excited about this new thing called "nonlinear editing."

So between the flashy sessions to Lucas's peers were hands-on events with working editors. Ron Diamond would probably never buy one of these $180,000 laserdisc-based get-ups, but he sat with me the way every other editor sat with me: *totally blown away.*

Seven years later I ran into Ron Diamond at the ShowBiz Expo in LA. He didn't believe that I remembered who he was (and after about 2000 demos and students, I was surprised myself), but I did. He had read NONLINEAR 1 and quite enjoyed it. In fact he had become a serious student of nonlinear things, in many ways due to the book.

When I completed the initial draft of *N3* early in 1994, I ran sections by my panel of experts on specific topics, did the requisite fact checking, and then pondered who to run the book by for a final once-over. I called Ron.

About three weeks later he returned the 320-page xeroxed monster to me with a cover letter: "Enjoyable read. Thanks so much." However, when I flipped through the pages I was floored: Ron had made notes *on every single page of the manuscript.* I spent the next two months reading his fine print and trying to improve the book with the end-user's thoughts in mind. By summer I had burned out.

More than a year later, as NONLINEAR 2's effective lifespan was running short, I picked up Ron's notes and began a page one rewrite. Hungry to have the undivided attention of other experts in the field, now more than 5 years since the original book was penned, I called upon my long-time friend Derek McCants — introduced to me by nonlinear bwana Kenneth Yas at The Post Group — and now not only an accomplished editor but an instructor of things Avid and nonlinear. Derek agreed to be kidnapped for an intensive weekend in Santa Cruz; to go page by page through N3 drafts and revise, remove, or replace each segment as necessary. This we did only a month ago.

As Derek had long been a supporter of this project through its many editions, and was critical to the completion of N3; and as Ron had risen beyond the call of duty for the sheer joy of participating in a project that had at one time made a difference in his own career, I felt the very least I could do was formally introduce these fine individuals to the community of N3 and publicly acknowledge their special contributions to this ongoing body of work.

CONTRIBUTORS

Derek McCants is a professional nonlinear editor, consultant and teacher in Los Angeles. He has been involved with nonlinear editing since 1987, while assisting in the development, support and testing of the CMX 6000 laserdisc-based system. He has been involved in the teaching and support of many video systems both nationally and internationally for such manufacturers as Sony, Ampex, Panasonic, and Calaway, and has been an instructor for Weynand Training in Los Angeles since 1987. Derek has been a pioneering video editor since 1979 and over the years has been on the editorial staff of "A Current Affair", "The Today Show", "Lifestyles of the Rich and Famous", and "Hard Copy". He was senior editor of "The Crusaders" and has been acknowledged for his documentary work on PBS's "NOVA". His freelance editorial clients have included Paramount, GTE, Fox, AT&T, NYC's Museum of Modern Art, and Entertainment Tonight.

After having cut his first Avid project in 1989, Derek has recently become an Avid Certified Instructor, and has completed teaching numerous courses in LA and at the Maine International Film and Television Workshops.

Ron Diamond still fondly recalls his first editing job. In his early teens, he recorded the off-air soundtrack of an entire week's worth of Johnny Carson monologues and skits, and on a whim, spliced them together into a mock composite using scissors, adhesive tape, and a beat-up old reel-to-reel tape recorder his mother found at a garage sale. Ron went on to wear a variety of professional hats during the 1980's, including radio engineer and producer, technical director at Boston's CBS television affiliate, and video editor; and he credits some influential early mentors and colleagues in passing on an editorial aesthetic that has served him well. In 1992, he segued to Los Angeles, becoming an accomplished Avid editor on an array of long and short-form projects, syndicated and network, and in diverse genres including promo, documentary, corporate, news, and episodic television.

He also looks back on the first decade of the democratization of nonlinear with a hint of nostalgia, considering himself privileged to have witnessed its genesis firsthand at Lucasfilm, a long time ago.

C O N T E N T S

PREFACE TO THE THIRD EDITION

I entered the world of nonlinear editing in 1985. With no preconceived notions and no agenda, I observed the realities of film editing, videotape editing, and the new realm of nonlinear editing. It was clear to me then that the new alternative was a vastly superior methodology, and sooner or later, would revolutionize editing.

For the next five years, I joined a handful of missionaries representing a half dozen churches who set up shop in the jungles of Hollywood, converting the natives and trying desperately not to be slaughtered in the process. At NAB 1991 the digital nonlinear systems were creating as much a stir as the original systems had 5 years earlier; but the missionary work was not completed — in fact it was just beginning. It was from this environment that my book NONLINEAR was first produced.

Today, just about a decade from my adoption of nonlinear technologies, I have run the gamut from manufacturer, teacher, editor, consumer, to designer; and I have to say that while NAB '95 revealed that just about every editing manufacturer was releasing their own nonlinear system (or absorbing a company who had one) it is about as *revolutionary* as a good old linear keyboard system was in 1985.

This volume will hopefully fulfill the interested spectator's appetite for information on nonlinear systems. The brands of equipment and the companies who produce them are finally widespread and thus in many ways unimportant. Where once there was CMX, now there is Avid. The names change but the song remains the same. This book is updated, streamlined, and most importantly, useful. It is not a buyer's guide. **It is a field guide.**

So dog ear a few pages; flip around when some client has you confused or some manufacturer is speaking gibberish. The new topics in this edition — spanning more than 60 additional pages — mostly cover updates and advances in digital technologies, new standardized 3:2 definitions, an outline of my theories and observations, and a neccesarily revised section on systems.

As when I first wrote NONLINEAR, I have few corporate biases and no hidden agenda. I represent no manufacturers and don't wish to get you to purchase any specific products. My only goal is to create a community of consumers who care about editing, who have a healthy disdain for technology, and who are informed enough about the things that matter to be able to navigate the tricky waters of post-production in the digital age.

So hang on, enjoy the ride, thank you and good luck.

October 1995

PREFACE TO THE SECOND EDITION

Much to my surprise, a lot has happened since I started writing NONLINEAR. When electronic nonlinear editing began, computers were simply the *controllers* for a vast array of peripheral video equipment: switchers, videotape decks, laserdisc players.

And just when filmmakers started to get used to *those* computers, look what happened. Another revolution. All the video, effects, and titles are now *in* the computer instead of attached to it. And these aren't zillion dollar high tech university computers — these are desktop computers. These are Macs and PCs!

Today, almost anyone can be an editor. But not everyone can be a *really good* editor. Talent, more than technical know-how, should become the valued commodity. Just because everyone can type doesn't make everyone a novelist, right? The democratization of post-production will no doubt lead to some horrific displays of bad editing. But like the super 8 home movies of the 1950s, it will also introduce filmmaking to a new generation.

As in the first edition, NONLINEAR 2 is not about editing *systems*. It is about *editing*. Throughout, I have updated information where appropriate, and added new information based on readers' requests. New sections now cover topics like component and composite video, digital compression, and QuickTime.

Because of the new individuals now uncovering nonlinear editing in the form of desktop video, and the number of schools around the world that have been teaching linear tape editing for a decade with little modernization to their curricula, this book appears to have filled a void in more than just the high-end professional markets for which the project began.

The coming years will see the down-scaling (and down-pricing) of many of the professional editing systems and the simultaneous upscaling of the consumer computer products that handle video. The years may also show us radical changes in the post-production facility business. Facilities will always be needed for the really expensive things (like telecine, dubs, and high definition), but a single person can pretty much do all the editing, graphics and effects without leaving the living room. In some way, this was the original desire George Lucas had when he began his EditDroid project. He wanted to make the complex, labor-intensive work of post production considerably easier. We are seeing this now, less than a decade later.

But through these years, the fundamentals and principles of editing and editing systems have not and will not change. In coming years, this book will serve a dual purpose: teaching novices the basics of nonlinear post-production as well as serving to update professionals on the changes going on in their own industry.

For professionals more than anyone else, the pill of technology may be a bitter one. It may not be here yet, for everyone, but it is closer than many realize. The technological revolution the world is going through with HDTV, video telephones, fiber optic cable networks, and interactive media will all directly impact editing and post-production; at the most obvious level, it will demand better video compression techniques and better-looking digital images. For many people, it has already arrived.

As the changes come faster and faster, more individuals will discover the need to understand both nonlinear basics and the fundamentals of the new technologies.

June, 1992

PREFACE TO THE FIRST EDITION

At this year's National Association of Broadcasters show I met many videotape editors who thought the new digital editing systems they saw were the first nonlinear products. I met film editors who didn't know the difference between hard disks and laserdiscs. As I walked the exhibit floor, it became apparent that editing and editing systems were about to get . . . confusing . . . if they weren't already.

For over six years I have been actively involved with nonlinear editing, and only today it seems the editing community is sitting up and saying, "Hey, look at this nonlinear thing . . ."

If you are finally discovering nonlinear editing (or only now find it ready for *you*), you are probably in need of a perspective on these systems: where they came from, how they do what they do, how to work them, and which is the right one for you. So where can you turn for help?

No person in the business can give you an objective overview. Although I promise to try, it probably won't be absolutely evenhanded. No one is unbiased and few know all the systems well.

Not the *trade press* — they have magazines to sell and advertisers to please.

Not the *manufacturers* — they really don't know much about *any* editing systems other than the ones they make (and sometimes don't know that much about their own).

Facilities? They tend to get informed once they've owned a piece of equipment for awhile (often knowing it better than the manufacturer) but will have only a very selective knowledge about any other systems.

Producers? Their principle concern, and rightfully so, is cost. Their judgments are important, but their bias is toward economic and logistic features more than editing functionality.

Editors? Most who edit nonlinearly have fallen in love with the first system they learned and rarely want to venture out much farther. There is a growing body of editors who started on film, experimented with videotape (but generally rejected it), and then discovered nonlinear video. Of this group, most have edited on two or three different systems at different times, giving them a somewhat diverse knowledge of system functionality. These are good people to talk with. You will find their recommendations divided between the system they *like* the most and the system that most productions want them to use.

No system is perfect. Everybody wants different things and every system offers different things. The "right" features are in at least half a dozen different editing systems but no one system has them all. Today there are at least twelve significant manufacturers of nonlinear electronic editing systems. By the time you read this, there may be twelve others. Though the specifics of each system may be different, there is much background that is fundamental to all of them.

This book does not presume to teach anything about the aesthetics of editing. It arose simply from a need to demystify the myriad editing equipment choices an editor has today, to bring to editors a unified terminology and a basis for learning more about systems not yet developed. Technology happens whether we like it or not. Consequently, this book was designed for 1) film editors who need to understand more about the world of "electronic film" editing — to develop some understanding of the techniques, vocabulary, and technologies of the video world; 2) videotape editors who want to investigate the stylistic power and history of nonlinear editing; and 3) producers and directors who are baffled by their electronic options.

As the decade progresses, videotape and film editors will be more and more in the same ship — learning new ways from each other about editing in the future. I hope this book will begin the foundation for that new community.

November, 1991

INTRODUCTION

No book can replace actually working in post-production and with nonlinear editing. The purpose of this book is to familiarize the reader with fundamental concepts associated with nonlinear post production. It should be read in conjunction with class training or real-world experience.

For learning how to operate any of the available nonlinear electronic systems, manufacturers provide classes, the Editors Guild supports training, and various universities offer special programs. However, technical background is often passed over in the classroom setting for a more direct hands-on application of knowledge.

NONLINEAR is not meant to be read cover-to-cover like a novel. It is completely random-access and nonlinear. The reader should feel free to peruse sections and find topics that are of particular interest.

Many topics are repeated in different chapters and in slightly different ways. Film, video and computer people often have vastly different backgrounds, and some will need more explanations on certain topics than others. NONLINEAR assumes the reader has only a rudimentary knowledge of either film, video or computer concepts and terminology.

This "primer" is neither intended to teach you how to edit on any particular system nor to convince you which system is "best." It will discuss the history of the nonlinear systems and give an overview on how they work. When appropriate, I will offer my opinions on the good and the bad, hopefully without proselytizing and without malice.

Additional Notes

On Terminology

There is no uniform terminology in the developing hybrid of film and videotape editing. Since individuals approach this world from either a film or video (or computer) background, the vocabularies often reflect one of these worlds but not the union of them. The problem would not exist if we had a central editing school teaching one set of terms; but since there is none, perhaps this handbook of nonlinear editing can help achieve the needed standardization.

Thus, it is suggested that the terminology found here be adopted by, and disseminated to, all concerned with nonlinear editing. These definitions and spellings are not necessarily better than others used elsewhere, but are those generally accepted. After all, it is not so important which term is used, but that the term is understood by everyone.

On Specific Styles

For our purposes, this book will use the following conventions: *online* and *offline*, meaning editing types, are one word, as is *nonlinear*. *Pre-* and *post-production* are hyphenated. *On-line*, also with a hyphen, describes the state of being connected to a system or network. And regrettably, *broadcast quality* will often be used to describe the image quality associated with traditional television broadcasts, comparable to CCIR-601 uncompressed video, until a better objective term is devised.

On Systems

To make this book as useful as possible, actual systems are named and some technical specifications are printed. It should be recognized, however, that these systems are always changing, upgrading, modifying their pricing, and releasing surprise features to anxious trade-show audiences. Through most of the book, great care is taken to describe features in the most general way and not specifically as a description of a particular product. Although some editing features are patented, most are general enough to appear in many systems.

Only parts of a couple chapters actually name individual nonlinear products, the prevailing ones at the time of this writing. All are certain to change over the months and years following this publication. The background history is culled from dozens of interviews and countless articles with key players in the nonlinear game — inventors, company presidents, facility staff, and editors. Although it often reflects many of their biases, it portrays an accurate and fair view of the evolution of these products.

CHAPTER 1

FIRST THINGS

WHAT IS LINEAR EDITING?

You're having a party. Your job, starting early that morning, is to prepare the "party tape" — a cassette of music that you can pop in the audio cassette deck and that will run throughout the evening. You collect up your favorite record albums (the *source* for your party tape), choose your favorite songs, and start recording. What do you do first?

Insert a cassette tape in the tape deck, cue it up to the beginning, and then load up the first album onto the turntable. Let's say it is Madonna. *Like a Virgin*. Press *record*. Drop the needle. You're rolling.

Four minutes and 23 seconds later, the song ends. You press stop at just the right time, and then unload the record. One down, 28 to go. Load the second album, Ritchie Valens. Re-cue the tape after the Madonna song, and prepare to drop the needle at the beginning of *La Bamba*.

Whoops. You missed. Stop recording and re-cue everything. On the second try you get it right. While it is recording you run to the kitchen to put beer in the fridge. By the time you get back into the living room, *La Bamba* has ended and the next song on the album has already started.

No problem. Stop the recording. Stop the album. Rewind the tape to cue it at the end of the second song.

By afternoon you have a terrific party tape. The *list* of songs, and their durations, looks like this:

Like a Virgin	4:23
La Bamba	2:04
Bad Moon Rising	2:18
No Woman No Cry	4:03
When I'm Sixty-Four	2:36
Blinded Me With Science	3:42
Stairway to Heaven	8:12
(plus 21 others)	

What you have done, in a very real way, is perform an *offline/online edit*. A traditional, *linear* (albeit audio-only) edit session. That's what a party tape is. Linear editing is simple and very understandable. LP records were your *sources*, the *edit list* is printed above, and the party tape is your *edited master*.

SO WHAT'S WRONG WITH LINEAR EDITING?

Nothing, really. Except this: let's pretend that you are checking over your party tape an hour or so before everybody arrives, and you decide that you'd rather not use the second song, *La Bamba*. It used to be your date's favorite song with Somebody Else. Now what? RE-EDIT.

All you have to do is record a new song over the old one. No problem. You pick Springsteen's *Born to Run*, but find that it is almost 5 minutes long. If you start at the end of Madonna, it will roll over *La Bamba* (2:04) and the beginning of the next song, *Bad Moon Rising*. So that won't work. You really like Chopin's Minute Waltz, but, of course, it is only a minute long. You'd still have 1:04 of *La Bamba* left over, hanging out the end. What do you do?

Short of finding a new song that is *EXACTLY THE SAME LENGTH AS THE OLD SONG* (which is extremely unlikely), you have only two options: record the new song, and then record all the following songs *again*, from their original albums; or copy (dub) the first tape onto a *second* tape, record the new song in the right place on the first tape, then re-record the dub back onto the first tape in the right place.

The first option will take all day because it means re-locating and re-cueing every song. The second option is much faster, but the music will lose a couple generations of quality from all the re-recording. For your party, you probably will opt to skip all of this, turn the volume down low, and hope your date doesn't notice — because LINEAR re-editing is sometimes more trouble than it is worth.

HOW CAN NONLINEAR EDITING HELP?

If the party tape had been created on a nonlinear editing system, like a multi-disc CD player, you would only re-program the order of the songs you wanted, skip *La Bamba*, add other songs, do whatever you wanted. Ten CDs, the number of discs you can **load** at once, determines how much music you can listen to without using any tape, or before needing to re-load. In fact, if all the music you want is in your load, you don't even need to make a party tape — you can play the music "live." It would sound better and if the mood of the party changed it would be easy to modify the selections . . . .

Nonlinear editing is a method for editing that allows editors to work in any order or style they want. Unlike traditional (linear) videotape editing, an editor can insert frames into an edited sequence, shoving everything that follows farther down in time, or remove frames and have the following material slide up. Unlike an edited master tape in video editing, each shot is not locked to time — it exists only as a "source in" and "source out."

Like editing film, a nonlinear work can be roughed together, and then slowly *honed* and *sculpted* into a final cut. In linear videotape editing, however, this isn't really a practical way to work. And since it can be so much trouble making certain changes, editors and clients are sometimes forced to skip those changes, and ultimately the final product suffers.

At the utopian ideal of nonlinear editing, one would have instant access to ALL source footage, and there would be no drawback in modifying ANY particular edit — whether it be the second shot in a list or the last. Film editing is nonlinear. Of course, accessing source material is not necessarily quick, and the handling of trims and coding of dailies is relatively laborious. For some kinds of projects, particularly those with relatively small quantities of dailies, this ideal of electronic nonlinear exists today. For other projects, larger projects with a lot of source material, true nonlinearity is still only approximated, but to acceptable limits. Total nonlinearity may not always be better.

Since videotape is *linear* by its nature, the longer an editing system can keep the edited master from being recorded onto a piece of videotape but still show it to you (the more shots it can *preview*) the *more nonlinear* it is. Some systems can only preview a few shots, others can preview short sequences, and still others can preview virtually unlimited scenes before necessitating any record tape. In other words, nonlinearity is 1) purely a function of the way the editing system schedules its "look-ahead" previews, and 2) the dynamics of the source machines' access speeds.

Related to this ability, but not defining it, is the amount of unedited source material that can be loaded up at one time on the system, usually referred to as a "load." In general, the larger the load, the more material that can be previewed without having to record on videotape. Some systems have physical limitations as to how much source can be loaded at once; others have virtually no physical upper limit, but significant economic or practical limits. Often these are just as debilitating.

There are many features of nonlinear editing that are not implicit in the word "nonlinear," which nevertheless are embodied in this style of editing. These include source material logsheets for fast access to dailies, 30fps and 24fps output possibilities using film/tape relational databases, and graphical human interfaces. These are all discussed in Chapter 4, Editing System Primitives.

B U Z Z W O R D S

NONLINEAR   FRAME ACCURATE   FILM-STYLE
RANDOM ACCESS   REAL-TIME PREVIEW
DIGITAL   DESKTOP   BROADCAST-QUALITY

If you've been reading the trade magazines on video or editing any time in the past half-dozen years, you've seen the "buzzwords" listed above over and over. They appear in informative articles about electronic nonlinear editing systems and in the advertisements for them. Sales people, especially, string the words together to make their system sound particularly powerful and impressive. Unfortunately, the words have been used and misused so often that they are on the verge of becoming almost meaningless adspeak.

Do you really know what they mean? Once you understand these terms, you will be better able to understand what a magazine or manufacturer has to say. Identifying the best system for your needs still won't be easy, but you will be off to a good start.

Oddly enough, these terms not only can be used to describe an editing system, but they are generally the *defining* characteristics of this type of product. So, while the words as used may well be true, they don't necessarily differentiate among the various products.

"NONLINEAR"

As you now know, traditional videotape editing is, by definition, *linear*. It means that first you make the first edit, commit to it, then make the second edit, and so on, starting at the beginning of your project and ending at the end. Changing any edit that is not your current edit is somewhat difficult, and the farther back toward the beginning of the list you go, the more trouble you may have.

"Nonlinear" is the key word describing the new computerized editing systems that are designed for cutting film, or at least cutting in a film-like style, and needing to record to videotape rarely. Some may be "more" nonlinear than others, meaning they have a greater degree of "preview-ability," but that does not detract from the fact that all approach the nonlinear ideal. As it is a somewhat measurable feature (more so than "random access" or "frame accurate," for example), "nonlinear" is a good term for the general category of editing systems discussed here. It serves to separate these technologies from the rather wide field of *linear* electronic systems that have been in existence (and evolving) for 20 years.

By doing away with recording the edited master to videotape, any editing system, regardless of its source capacity or type of media, its human interface, or its total speed in accessing, can be nonlinear.

"FILM-STYLE"

"Film-style" editing, at its simplest definition, is identical to "nonlinear editing." It is the style for editing film. (Similarly, "video-style" means linear.) However, "film-style" is a term that has been applied to systems that *softened* the technological, numerical, and cold interface that videotape editors used for something more "humane."

The first general characteristic of a film-style system is that it utilizes a computer monitor with a bit-map graphic display (as opposed to simpler computer screens with only ASCII characters). Film-style systems have adopted various kinds of fancy graphics and icons that are intended to make the computer more user-friendly for film editors. Pictures of film. Icons of spools turning and synchronizers spinning. Trashbins and index cards. Scissors for cursors.

Film-style is a marketing term for *film-editor-friendly interface*. These electronic editing systems were sometimes the first computers older film editors had ever seen. In the first years of electronic film editing in Hollywood, editors who had been editing television and movies on *film* — sometimes for 30 or 40 *years* — were asked to adapt to these new computers in weeks or sometimes *days*. It was of critical importance that film editors find them easy to learn and simple to use. In other words, they had to be a lot like editing on film.

There are many ways in which a computerized system can be visually styled for film makers: It can have film icons. It can be script-based (displaying some kind of lined script as part of the interface). It can actually display film, giving you the ability to zoom so close to your material that you are actually watching individual film frames move by. All these are methods to *de-technify* computerized editing.

Regardless of the "film-style" of a computer, film editors continue to miss the tactile nature of editing film. They enjoy both the nonlinearity and the physicality of working with celluloid. No matter how good electronic systems get, they are rapidly losing the physical nature of editing.

"FRAME ACCURATE"

This is a tough term. It has many meanings to many people and is more misleading than any other marketing phrase.

The term never existed in the world of film. Editing film *is* frame accurate. You might have asked, on occasion, if a negative cutter matched the workprint *frame accurately* . . . but that is another story.

The term showed up when editors began offline for videotape, and timecode was the key. If I am using a computer to control a piece of videotape, and I am looking at a picture on a monitor, I generally want to know "What frame am I looking at?" If I have striped timecode along my videotape, my computer will display that timecode on my screen.

BUT, there are different kinds of timecode and each has different features and costs. *Vertical Interval Timecode* (VITC) is extremely accurate, but it used to be relatively expensive. No matter what frame I am on, whether I slowly shuttle forward then backward, then fast, then slow, it always knows the exact timecode of the frame I see on the monitor. *Longitudinal Timecode* (LTC) is less expensive and potentially less accurate. With LTC it is possible to confuse the computer by marking an edit point after shuttling around erratically, in non-play speeds. If for any reason the machine misreads the LTC, you may try to make an edit on frame 3, for example, but the computer might actually record the edit on frame 4 or 5. Your machine and editing system are thus *NOT FRAME ACCURATE*.

This is the original meaning of frame accurate: *the computer knows the video timecode of the frame that you edit on*. Always. Accurately.

Some nonlinear systems use LTC as their cost-effective timecode. These systems, just like any videotape system using LTC, are potentially not frame accurate. Digital- and videodisc-based systems, on the other hand, are extremely frame accurate. Computers and disc players are very good at cuing to and isolating a particular frame identification. Unfortunately . . .

THIS IS GENERALLY NOT WHAT PEOPLE MEAN WHEN THEY SAY "FRAME ACCURATE" WITH REGARD TO NONLINEAR SYSTEMS.

Nonlinear editing systems were created to give film editors the power of modern videotape editing, but with a style that they were used to. The process was thus highly creative, like film editing, but like videotape editing involved no head/tail trims and allowed for faster access to material.

Most nonlinear projects start with a film shoot, and although they might air on television, they often want to finish on film as well. Since film speed in the USA is 24 frames per second (fps), and videotape speed here is 30fps, there is a problem of mathematics when editing something on videotape that began and must end on film. Without going into detail here, whether or not a system accurately does this math — allowing a cut on videotape to "frame accurately" conform back to film — is the crux of this term.

There are two principal kinds of accuracy: *visual accuracy* and *temporal accuracy*. If a cut is visually accurate, it is frame per frame identical to the cut the editor created on the editing system, but it is still possible that the ultimate timings of cuts are off. If a cut is temporally accurate, it will have the same duration as the electronic cut, but it will only be visually accurate to a point; isolated frames can be added or lost to keep the timings in sync.

It is impossible, in the 24/30fps world, for an electronic edit to be identical to both the film cut *and* the video cut. Only one can be truly identical, and the other will be an extremely close approximation, based on a mathematical formula. Type of accuracy and method of conversion varies from system to system. All systems are *frame accurate*, depending on your own particular set of defining parameters.

"RANDOM ACCESS"

This is an odd term that is unusually common. All editing is random access — that's what editing is. You access THIS scene, then you go to THAT scene . . . then look around for some other scene . . . .

What is generally meant by "random access" is that there is *quick* access to randomly selected points in the dailies and/or the master. If you want to go from the beginning to the end of a particular scene, it's really not random access if you have to shuttle through the middle. It *is* random access if you just "pop" there. Compact Discs, videodiscs, and computer disks all offer pretty random access to source material. As long as dailies or created edits are NOT recorded to videotape, that system is going to be indisputably random access. If you can edit scene 1, then edit scene 20, then quickly jump back and change something on scene 1, that's random access. To be truly random access, like being truly nonlinear, you shouldn't suffer any inconvenience when moving to any location in your material.

Videotape is not very random access: but edit systems can achieve a *simulation* of random access by having multiple copies of the same material on several reels of tape. Then, one tape can be cued to the head and one to the middle and one to the end . . . and you can jump between them. Otherwise, an hour-long videotape can take many minutes to shuttle from the beginning to the end.

Videodiscs are very random access (very quick moving), although you generally need at least two identical copies for nonlinear editing. Often more. Still, it is enough faster than tape that you need considerably fewer copies for identical levels of performance. A 30-minute videodisc can jump from frame 1 to frame 54000 (the end) in less than 2 seconds.

Computers, with digital memory, are extremely random access. When your images and audio are digitized as part of computer memory, you have the ultimate in random access: RAM is virtually instantaneous.

For the ideal in random access, all source material should be available upon demand. Unfortunately, cost and practicality tend to limit source capacity as much as technical limitations of the various systems. Nonlinear systems, which all provide for a degree of random access considerably greater than other kinds of editing (traditional film or tape), are often referred to as "random access" or "RA" editing systems. Actually, any kind of electronic editing that does not use a single videotape per source can be called an RA system.

It is important to note that "nonlinearity," "film-style," and "random access" have *unique* meanings, and can only be used interchangeably in rare cases. In general, nonlinearity is a kind of editing, random access is a kind of cueing, and film-style is a kind of interface.

ONE DRAWBACK OF RANDOM ACCESS

With the advantages of powerful random access searches and nonlinear editing, there is also a noteworthy drawback.

Videotape and film editing both require editors to shuttle through material to locate dailies and cut sequences. Random access editing decreases the amount of shuttling that an editor must do. Although this sounds like an advantage (and certainly is) in time savings, it reduces the familiarity an editor has with the raw and cut material and could therefore actually adversely impact the quality of the final product. This is one criticism sometimes raised by experienced and inexperienced editors alike.

"REAL-TIME PREVIEW"

For the first many years of analog nonlinear editing, the way the editing sequence was viewed was through a "preview" or "real-time preview." This was a variation of a common video editing term which further began to define nonlinear systems.

A *preview* is a *rehearsal* of an edit prior to committing that edit to a record tape. If an editing system cues up one or more sources and then plays a sequence for you, that is a *preview*. If the computer controller is *smart* enough, while one shot is being presented to the viewer, all the remaining machines can begin shuttling and cuing to the next location needed for playing following shots. A series of machines can kind of "leap frog" while you watch them simulate a recorded master tape. This is a "look-ahead." This is how analog systems achieved nonlinearity.

Film editors have no equivalent concept for the preview — you cut something and you watch it. You don't like it? Change it. Now. Later. Whenever. But because videotape is linear, you want to watch an edit before you record it — because once it is recorded it is more difficult to modify. Watching a shot generally means watching it in the context of your edited master.

In linear videotape editing, there are three primary kinds of previews:

• BVB *(black, video, black)* — you see the shot you just edited as if it were between two pieces of black. You can judge how it looks or how long it is.

• VBV *(video, black, video)* — black between two existing shots in the master.

• VVV *(video, video, video)* — your new shot as if it were edited into your master. If you like it, you can record it and go on.

While these previews may seem phenomenally limited today, for a long time they were the only way to watch an edit before it was recorded to tape. And while the term "real-time preview" was used often, "look-ahead" preview was actually more appropriate and made more sense. What kind of preview is not in real time?

True nonlinear editing would mean not needing ANY record tape As soon as a shot or sequence is on tape, changing one shot means re-recording following shots — and you are back in the linear world.

The faster a source machine can shuttle and cue, the fewer copies of each reel are necessary. Digital systems, although they are not recording to tape, do not "look-ahead" to preview because they do not shuttle video transport decks for editing. Thus, with the advent of digital nonlinear editing, the term real-time preview and look-ahead preview became somewhat obsolete. But after a few years, a relatively new meaning took over.

The traditional real-time preview concerns what this book calls "horizontal" nonlinear systems and horizontal nonlinearity. This means that the separate shots are presented to the editor as if they were a single cohesive unit — in other words, as if they were recorded onto tape.

However, the digital systems have new functionality in terms of layers of video tracks. Traditionally, when multiple layers of video (and effects) were built and composited together in analog systems the editor saw the results of her work as she performed them. But digital effects are enormously complex for computers to handle, and often the editor must wait to see the results of compositing until the system chews on the data in each layer and puts them together. When modern systems preview these vertical layers in real time, this book refers to it as vertical nonlinearity.

Many of today's digital nonlinear systems are both horizontally and vertically nonlinear. In other words, they often offer some degree of effects compositing functions and previews (from little to mind-blowing). Since all digital nonlinear systems posses real-time horizontal previews, the term *real-time preview* today is generally referring to a system's capacity to preview composited effects in layers of more than one video track.

"DIGITAL"

Since the beginning of the 1990s, and the release of the first EMC2 and Avid nonlinear systems, a new word has risen up to replace many of the familiar catch-phrases of the industry — "digital". The term refers to the fact that with the new systems, source material is stored digitally on the random-access media, thus differentiating these systems from the analog tape-based and laserdisc-based progenitors.

Digital is *not* synonymous with "nonlinear." There have been ample nonlinear systems that were not digital; and there are ample digital systems that are not nonlinear. The original nonlinear systems achieved nonlinearity via multiple copies of analog (tape or disc) source material under the control of a central computer. Today's digital nonlinear systems simply remove the need for multiple copies of source material due to the nature and speeds of digital data manipulation in modern computers. On the other hand, there are currently many digital systems and equipment that are not nonlinear editors; "digital" only refers to the format of the images being manipulated. Linear video editing systems, for example, may use digital tape formats in the post-production process; or compositing systems may digitize high-quality video; all may fairly be described as "digital" systems.

It has been through steady advances in digital storage and computer power economics, along with contemporaneous improvements in digital compression techniques that have yielded the switch from editing systems with analog sources to digital ones.

"DESKTOP"

In 1985 Aldus released a software product called "PageMaker" that brought professional publishing powers to a person with a Macintosh computer. In doing so, Aldus coined the term "desktop publishing" to describe their new revolution.

When the Macintosh was released in 1984 it included a new kind of bit-mapped display (new for personal computers, anyway) with icons instead of the familiar computer text CRT; Apple officially termed this interface "the Desktop." Aldus PageMaker was making a kind of double reference with their new term. In the first place, the Aldus software literally ran on the Macintosh "desktop," and since the product was designed for and released on the Mac, it was quite realistic to call PageMaker a *"desktop* publishing system." In the second place, but of greater importance today, they were truly taking a set of tasks and procedures — from typesetting to page layout — and integrating them into a single tool that could be performed on a personal computer, in essence, bringing publishing to a single "desktop." In this sense, the word has come to imply that a personal computer sits on your desk and is run by a single individual.

Since the revolution in publishing brought about by PageMaker and other desktop publishing tools, people have sought to expand on the term's meaning by using it as a suffix for other industries: in particular, desktop video.

Today the word is used for its strong emotional impact: it depicts a system that runs on a personal computer, and is thus smaller, less expensive, and easier to use than its non-desktop counterparts. Ironically, as personal computers have evolved, a "desktop" system for either publishing or video can very often involve many pieces of equipment (that have trouble fitting on all but the biggest desks); they can be reasonably expensive and generally difficult to use.

Today the true meaning of the *desktop*, now that it no longer implies the exclusive Macintosh domain, involves a software plus hardware set-up that runs on off-the-shelf personal computers rather than dedicated platforms; it is a tool that while perhaps professional, is still more within the grasp of the average user; and most importantly, it is a break from the production methodologies of the previous era, where tasks that had once been diverse are now integrated (while not entirely in software, at least in hardware).

Unfortunately, the *desktop* is often a type of diminutive term for professional equipment, but ultimately tends to relate more to individual empowerment and liberation.

"BROADCAST QUALITY"

Since video information was first broadcast — that is, transmitted through the air to receivers located in your house — it has well been understood that the *strength*, and thus the *quality* of that broadcast signal degraded substantially by the time it got to you. Consequently it was very important for those doing the broadcasting to have a very strict set of engineering guidelines to maintain the technical quality of the shows you see on TV. These strict rules were based on the parameters of the standard video waveform: signal to noise ratios, amplitudes of colors, sync and blanking (don't ask), *etc.* While the exact guidelines are perhaps unimportant here, what should be noted is that the term "broadcast quality" is not an arbitrary criterion for images and sound — it is a very well-documented *objective* series of limits to insure that the shows you get are designed correctly for broadcast (and by the way, looking good was kind of secondary). For almost forty years, the SMPTE-set rules, adopted by the FCC, defined what you could and could not broadcast.

The relationship of image quality to broadcast quality is incidental. So while broadcasters were concerned with getting tapes with broadcast-quality video signals on them, it was the networks, stations (and advertisers) that were most concerned about the actual quality of the images. Networks each came up with their own image quality criterion; but in time, limitations on image quality were dropped. It is marketing-speak to use the term "broadcast quality" to relate to image quality. Still, common usage tends to toss around the *image quality of broadcast-quality video* to actually define *broadcast-quality video*. This cannot be emphasized enough: broadcast quality is the quality of the **signal** and not the **image**.

WITH THAT HAVING BEEN STATED FOR THE RECORD . . .
The question arises: can you take a less-than-high-quality image and place it into a broadcast-quality signal? (yes) And what **is** the expected image quality of broadcast-quality video? (high).

Well, it used to be that if the picture quality was deemed *low*, it would not be considered for broadcast. Thus, video images had to be recorded on high-quality equipment in order to be appropriate for TV broadcasting. But the highest-end technology of the 1970s was often less able than today's camcorders to deliver this high-quality product. With the advent of Y-C camcorders (*e.g.* S-VHS or Hi-8) and then with the proliferation of digital video systems, the quality of video that the public would accept on TV seemed to be less a factor of the "image quality" *per se*, so much as the content. Frankly, if some consumer had a home video of a train wreck, the actual image quality certainly wasn't going to stop it from going on the air. Then with network show after show of consumers' home videos, the tolerance of the TV viewing public to lesser image qualities began to drop. Which brings us to digital images.

A frame of video is considered to have 525 lines of resolution, refreshing on the screen at 60 fields per second (30 frames per second). But only about 480 of those lines of resolution show up on the screen. In computer terms, an image of "full" resolution (the rough equivalent of the "broadcast resolution" criterion), is about 640 pixels across by 480 pixels down, with each pixel having one color out of a spectrum of 16 million (or 24 bits). When you work out the math, this gives you a frame of about 900K and a second of video (without audio) at about 27MB. Since 27MB/second is tough for many computers to handle, compression techniques are used to make the video "smaller." Of course, for the most part, making the video smaller also means degrading the image — making it look worse. But many viewers cannot distinguish the uncompressed 27MB/second-video from the same image compressed to 2MB/second. And one system's image quality with 2MB/second video (68K/frame) will likely look a little different from another's, depending on what the images are, the method of compression, and so forth; so even describing an image as a given "bytes per frame" is not necessarily a fair comparison. The equipment manufacturers are left with a quandary: how do you objectively describe a resolution? Even if it's good enough to broadcast, it is inappropriate to describe it as "broadcast quality". At first, they did anyway until the officials got mad. Now, there is a whole array of subjective terms that describe image qualities. "Online quality" is a sloppy substitute, roughly corresponding to the images you might see in a professional online bay. Comparisons to tape formats are slightly more concrete: "Betacam SP quality" and "S-VHS quality" are two examples of common subjective image resolutions. These are not optimal solutions to describe image qualities, but they are considered acceptable today as long as users understand the obvious limits and "softness" of the description.

As an end-user of a nonlinear system, simply realize that there are two ways to describe an image's quality: via subjective descriptions and via objective ones. "Broadcast-quality video" is an objective description, but of a video signal; as far as image quality goes, **there is simply no such thing as a required broadcast quality**. Subjective comparisons of image quality are always possible, and certainly are the norm in the marketing of digital video systems.

THE GROWTH OF ELECTRONIC NONLINEAR

Most electronic nonlinear editing systems can be found in videotape post production facilities. You might wonder why film equipment rental houses or more independents don't have this equipment, since its original design was for the editing of film projects.

The fact is that many of these systems are expensive (prices only recently have begun a dramatic decline). Though offline editing has never been a source of great income to post production facilities, it is often used to bring in business. If you do your offline at Facility X, then you will probably do your telecine there, your online, your effects, your audio sweetening . . . and all those jobs are particularly profitable. Consequently, facilities are able to be very flexible with offline rates.

The big post houses adopted these systems initially because they could be used for film projects. Before electronic nonlinear was introduced around 1985, almost all broadcast television programming originating on film was edited on film, just like little movies. They were edited and screened and *everythinged* on film, and right at the end, when the show was finished, they telecined the 1-hour show to 1" videotape for delivery to the networks.

Besides the speed and flexibility the new "electronic film" systems offered over traditional film editing, they were able to generate new blood for the post facility community.

Work that was formerly done in film labs, optical houses, and film equipment rental organizations was now all possible in the video facility. Pioneering work in this vein was done at facilities such as Pacific Video's (now Laser-Pacific) *Electronic Lab*, and The Post Group's *Film Unit*, both in Hollywood.

This is primarily why the first nonlinear editing systems found a secure home in Los Angeles, specifically in the larger facilities working on network television. Only later did the systems begin to get discovered for their uses in commercials (also edited on film) and theatrical films. Finally these systems moved into traditionally linear territory that had been marked for music video, multi-camera video, corporate and industrial clients.

WHY TELEVISION SHOWS CUT FILM

There are a number of reasons why a production shot on film to be broadcast on television might want to cut the original negative.

In October, 1986, the *New York Times* published an article on the new revolution in editing. It referred specifically to the Montage, the EditDroid, and the Ediflex. The author described the evolving world of high definition television, and its incompatibility with current broadcast video. Apparently, *film* was considered to be the only viable way to convert to high definition. The article acted as a catalyst that galvanized interest in electronic nonlinear.

The second significant reason for having a film version of a television program is for foreign distribution. Europe and many other broadcast markets do not use the American NTSC video standard. They use PAL. Where NTSC broadcasts 525 lines of video at 30 frames per second, PAL broadcasts 625 lines of video at 25 frames per second. Although there are a number of very good "standards conversions" that will take an NTSC master tape and transfer it into a PAL tape, it is argued that only an original telecine from cut negative to PAL video is of sufficient picture quality to please European audiences.

Whether it is or isn't good enough is not the issue. The fact is that cut negative is as close to a super-high-quality universal standard media as the world knows. As distribution standards change, film will always be a wonderful source for transfer. Consequently, the need for projects to have a film cut is of ongoing concern to networks and studios.

HIGH DEFINITION TELEVISION

High Definition Television (HDTV or "HiDef") began as a new production standard that was first developed in Japan in 1970 based on research done in the US in the 50s and 60s. By 1977 SMPTE established a group to evaluate this new media, and by 1988, they sanctioned an analog standard called 240M, based on the Japanese's 1125/60 format. Although analog formats (like 240M, NTSC and PAL) require a broad base of acceptance and standardization, digital formats (for a variety of technical reasons) do not need the same degree of universality. The 240M analog standard was determined to be too limiting for the future needs of entertainment and information technology.

Since 1988, computer technology has been advancing rapidly; by the early 1990s, a larger picture emerged — one where HDTV was only a very small part. Futurists see all video as being digital, allowing it to be manipulated and distributed more easily. When video is digital, it can be played on computers, stored on floppy disks, and sent over phone lines. Once you see the future as digital, with computers playing a key role, you must modify your definition of HD television to be simply the broadcasted version of this large digital future.

Thus the term "high definition television" or HDTV is very limiting; the more general term for all these formats (including HDTV) is simply *digital television* or *advanced television* (ATV).

Digital television impacts not only television broadcasting (HDTV), but has applications in the movie business, education and medicine, printing and publishing, photography, defense and manufacturing.

The basic elements of broadcasting in High Definition are as follows:

• A system that provides wide-screen pictures with at least 2X the resolution (or sharpness) of ordinary TV, and sound of CD-quality. Traditional (NTSC) broadcast video consists of 525 lines of video interlaced from two fields per frame, and fields refresh at a rate of 59.94 per second; HDTV broadcasts video at significantly higher resolution — about 1000 lines on the screen, with many more pixels per line. A HiDef broadcast has more than five times the visual information of NTSC.

The issues of HDTV are profoundly complex and impact many varied aspects of how we perceive the future. For example, there are *production* standards which must be determined, and these are different from *distribution* standards. Production standards have to do with the way HD is shot, recorded, and post-produced. Distribution standards have to do with the many many ways this visual and auditory information is delivered.

Delivering video is just what you think it is: you can deliver on a video cassette, you can get a laserdisc, you can receive it over the air on a TV. Delivery of digital information — including video — can be done with floppy disks or telephone lines. So although broadcasting is a significant part of traditional video distribution, it is only one small part of *digital* video distribution.

The broadcast standard (or *transmission* standard, which is the same thing) has to fit within constraints that have to do with the number of channels on the air, among other things. The Federal Communications Commission (FCC) is the US government organization that primarily deals with over-the-air broadcasts. Other distribution paths include coaxial cables, fiber optic cables, videodiscs, CD-ROM, videophones and microwaves. There is no way to adequately have a single distribution standard; all of these video formats are somewhat unique. There could, however, be a single worldwide digital *production* standard; but determining what this would be would require an international effort of artistic, scientific and political cooperation.

Like the PAL, NTSC, and SECAM transmission standards, developing a new US transmission standard for high definition is of a high priority in the evolution of digital television. It has already been determined that in the next decade or so, the US will phase out NTSC broadcasts. Televisions, as we know them, will disappear. The future will provide for boxes which are both televisions *and* computers: playing digital images and audio either received as signals over the air, or from many other sources.

Four different groups representing seven different organizations had been competing to develop, at great cost, their own particular broadcast standards. Each was slightly different from the others. Each had been proposed to the US Government for one to be selected as the US broadcast transmission standard. Rather than choose one, the Goverment decided in 1994 that the four groups should work together — in what is now called "the Grand Alliance"— to complete the standard. The Grand Alliance is comprised of AT&T, General Instrument Corp., MIT, Philips, the Sarnoff Research Center, Thomson and Zenith. The FCC has insisted that their new standard must allow for a simulcast broadcast — where every TV station gets two channels (the current NTSC one and a HiDef one). This way, old sets can receive HiDef programs (in NTSC), and new sets can watch NTSC *and* HiDef programs.

Why is this such a big deal? Because there are 93 million homes with one or more NTSC television sets in the US today. That translates into a lot of new television receivers that might be replaced — with new kinds of television-computers ("teleputers") that take signals from all kinds of digital sources.

Once you go about inventing a new TV standard, there are many things you want to take into consideration. For example, no one wanted all the US TV sets to become obsolete, so the new standard had to deal with simulcast. Also many people wanted the system to be compatible with both film AND computers. Film is of extremely high resolution and of a different aspect ratio from TV. Computers use progressive scanning at their own frame rates. And of course, you want to make the HiDef images as good as they can be. Why go to HiDef if you don't get as much out of it as you can? The aspect ratio for HiDef — the shape of the image on the screen — will no longer be like NTSC's 4:3 (kind of squarish); HiDef is 16:9 (wider, like a movie screen).

To meld with our view of the future, the US has already decided that the standard must be digital (thus two analog contestants, including the NHK 1125/60 standard, were eliminated from the original HDTV competition). The issues that the Grand Alliance has formalized are many: they've chosen to use *both* interlaced fields like TV and progressively scanned frames like a computer, using the MPEG-2 compression scheme, digital 44.1KHz, 16-bit surround sound with 5.1 channels, and many other technical preferences. As you can see, the issue is complex.

The universal production standard will likely be unlike the transmission standards. One proposed approach involves digital RGB video with 1152 horizontal lines, progressive scanning, square pixels, and 72 frames per second. This creates a digital image of higher quality than any distribution options. This special RGB video can be encoded into a high-definition broadcast in the same way that RGB video today is broadcast using NTSC or PAL standards. The 72fps rate is an even multiple of film's 24fps allowing for a natural compatibility with standard film production. These are the types of issues which must be addressed in developing the ideal universal production standard.

One novel application associated with HiDef technology involves its use in the electronic post-production of film projects. Video today is only used as an intermediate working media to aid in the posting of film, primarily due to its relatively poor image quality. However, it is possible to transfer film to *high definition* video as if it were traditional video, post produce the project electronically (linear or nonlinear), and rather than *conform* film, just *transfer* the HiDef image to film. It is argued that the quality is an acceptable alternative to actual negative, especially when many films have enormous distribution in the video domain (cassettes, laserdiscs, TV, *etc.*). Consequently, film projects can benefit in time and cost savings from electronic post-production in terms of audio, titles, effects, and editing alternatives.

---

*For reference, film grain resolution for original negative is considered to be roughly the equivalent of about 3000 to 4000 horizontal lines at 30 fps. By the time film is projected in a theater, however, it is considered to have about 1000 lines.

STANDARDS CONVERSION

Until recently, television that was shot on film was edited on film. An NTSC telecine to videotape of the cut film was typical in this country, and for foreign audiences, a similar PAL telecine was done.

When film programming began editing on videotape, there was no edited cut on film, and the telecine to make broadcast tapes was unneccesary. The problem that remained was how to distribute a foreign PAL release if the only existing cut was on NTSC videotape. The primary available option was to transfer the NTSC master directly to PAL.

In the early 1980s, the television show *Dallas* caused some stir among European viewers. They wrote to their newspapers and television stations to complain that the image quality of the show was unacceptable. Apparently, the transfer from an American NTSC master videotape to PAL yielded a poor-quality master.

What had happened was this: a negative, telecined to 1" NTSC tape, was edited in offline, then the source 1"s were assembled to a master 1" of the show, making a 2nd generation copy. Usually the master tape goes through an additional tape-to-tape color correction, producing a 3rd generation tape as the NTSC master. This master goes through an NTSC-PAL transfer, tape-to-tape, creating a 4th generation PAL-master. And this PAL-master is subject to all the color and motion artifacts that might be created with taking a 30fps, 525 line source and producing a 25fps, 625 line master. It is understandably difficult to take a lower-quality image and convert it to a higher quality image.

Clearly, a cut negative generated from the offline would provide for a direct PAL-telecine, 1st generation from the negative, with none of the associated color and motion problems.

But the science of standards conversion has begun to change this. Although technologies are not perfected, many post facilities (and their customers) claim that their tape-to-tape standards conversion from NTSC to PAL provide as good or better quality PAL masters as a PAL-telecine. New techniques involve sophisticated digital video technologies, along with decoding systems for undoing the 3:2 pulldown that originally stretched the 24fps film to 30fps video. Artifacts, such as "judder" and "smear," are created in the 25fps world of PAL from the 3:2 pulldown; however, techniques that involve interpolation using "motion vectors" look at the changes in pixels from frame to frame, then estimate future locations of those pixels, and thus can reduce motion artifacts.

WHEN IS NONLINEAR EDITING APPROPRIATE?

While nonlinear editing seems too good to be true, there are still some cases where it is not yet the optimal editorial option.

Nonlinear editing shines brightest when it is time for the inevitable revisions. Moving shots around, shortening or lengthening shots in the middle of a show. In fact, in some ways, there is no type of project that couldn't benefit from nonlinear editing. However, there are elements essential to the process of working in the digital nonlinear domain, and some of these elements are substantially problematic to certain types of productions.

The key drawback in the current digital nonlinear process is in digitizing. For every minute of material that is to be used as source material for a nonlinear editing session, there is a minute or more of digitizing that is required beforehand. While this digitizing time is inconsequential in a feature film that may be in post-production for months, it can be devastating to any project that requires quick turn-around.

Take, for example, the news (or other timely programming). In spite of the drawbacks to linear editing, it is still often faster to pop a videotape into a deck, edit linearly (tape to tape) and walk away with a good edited master tape — sometimes dashing right to "air" (remember *Broadcast News*?) than it is to digitize a long tape or even segments of source material, then cut nonlinearly, then output digital video back to tape. Even if you digitize on-the-fly and don't have to log a videotape, there is often a cost savings in doing it the old-fashioned way.

This is also true for programs or projects with very few or very pre-defined edits. Another case may be certain newstories, promos or anything with a single uninterupted narritive. You generally must do a little analysis of the type of product you want to produce, what you must start with, and the costs of traditional methods. Nonlinear may be more fun, but it can be more hassle and more money.

Another point is that videotape and film are reasonably universal formats: that is, you can pull your tapes out of a session, take a floppy of the EDL, and go to another city and find another edit bay and get back to work. Digital files are not always compatible across system lines: you can't necessarily take your work out of an EMC and stick it into the first D-Vision or EditBox that you find. Doing so might often require redigitizing the source material into the new system. If moving around might be a real possibility, and dependence on a particular brand of system might pose a problem, nonlinear editing may not be the best option.

Projects with many, many effects are another potential problem. While nonlinear editing still could help, it is possible that the rendering time on a system would prove a great hindrance. Or more likely, a more "vertical" effects building and compositing system would be able to handle the source and output requirements better than a "horizontal" nonlinear system. Network TV commercials and promos with many effects may fall into this category at times.

And finally, projects that edit sporadically, with frequent starts and stops, may find the process of digitizing and archiving and re-digitizing, *etc.* more trouble than its worth. Truly, it depends on both the project and the particular nonlinear system you want to use.

Each year, advancements to systems and the post process — including digital cameras, fiber optics, ISDN communication lines, the OMF Interchange, OpenDML and new specialized systems for specific applications — will further streamline the usage of nonlinear systems. Consequently, limitations will need to be evaluated in the context of the features present at the time of a user's need. The myriad benefits of the new wave are undeniable, but each project always needs to determine the best method for it at the time; and sometimes, the answer may not be a nonlinear system.

SHOULD FEATURE FILMS ABANDON FILM EDITING?

Even as nonlinear editing has made dramatic inroads for feature film work, and almost all major directors and editors have given it a shot, it is not always clear if the digital nonlinear alternatives to film are ideal.

Film works. Everyone knows how to use it. It's the same every time. It has extremely high (the best) image quality. It is easy to take into the screening room and watch *large*. It's a universal format. And it's a known quantity.

So why switch?

To begin with, as anyone who has witnessed a nonlinear system in action can tell you, it is the creative power these systems offer. Changing edits is simple. Creating multiple versions is effortless. Nonlinear editing offers the ultimate in artistic freedom.

Secondly, electronic systems *can* be much faster than film. They can increase editing *speed* dramatically — some say 10% and some say 70%. But still, you can only edit SO fast; creative decisions still take time. However, changes and fixes can be rapidly tried and accepted or rejected, and as they say, time is money. A month saved in editorial means a month sooner the film is released and a month less *interest* paid on a huge loan (3 million, 10 million, 30 million). At 10% annual interest, a month saved on a 10 million dollar loan is more than $80,000 — not to mention associated labor and overhead savings.

On the minus side, projects usually expand to fill the time allotted. If a producer does not force saving time, it probably won't happen. In fact, the increased flexibility a nonlinear system allows could actually make a project take *longer*, if allowed to get out of hand. Increased flexibility translates into infinite variations and possibilities — a director and editor can literally be paralyzed by choice.

At a technical level there are only a few serious drawbacks to the electronic edit of a theatrical feature. The most important concerns the need to screen the film in a theater. The theatrical screening of dailies requires the printing of the film, syncing and coding by an assistant, all prior to telecine. The screening of cut reels requires that completed edits must be output for workprint assembly by assistants — the manual conformation of each reel requires from a few hours to perhaps the better part of a day. But both these tasks are reasonable; it is only when changes are made to the workprint on the electronic system, and the changes need to be screened, that the electronic process becomes unruly— this re-conforming process requires a good change list (not all nonlinear systems have them) and is still generally tedious. So until picture can be easily screened in high quality directly from the editing system, there will still be filmmakers who balk at the move into digital theatrical post-production.

CHAPTER 2

BACKGROUND

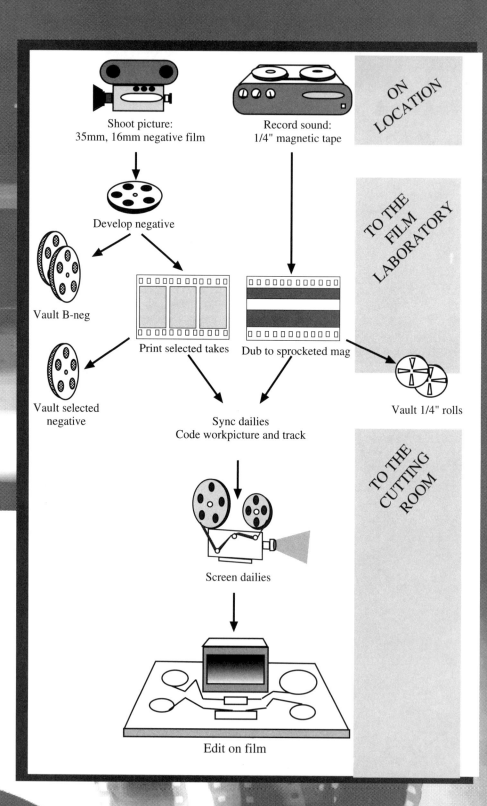

Shoot picture:
35mm, 16mm negative film

Record sound:
1/4" magnetic tape

ON LOCATION

Develop negative

Vault B-neg

Print selected takes

Dub to sprocketed mag

Vault selected negative

TO THE FILM LABORATORY

Vault 1/4" rolls

Sync dailies
Code workpicture and track

TO THE CUTTING ROOM

Screen dailies

Edit on film

NOVICE'S INTRODUCTION
IF YOU HAVE EVER WORKED ON A PRODUCTION. . .
DO NOT READ THIS

In the early years, nonlinear editors were converting from either a film background or a video background. But more and more, people are getting into nonlinear work with absolutely *no* background in editing. While this book would not presume to teach the novice how to edit, it has been suggested that a small explanation of the production and editing domains would be helpful. As promised, it will start at the very, very beginning...

Let's say you're shooting and editing a movie. A big movie. Lots of movie stars. You know how when you watch it on the big screen, it all seems to happen in real time. When you're shooting the movie, it doesn't.

So. You and I are in a scene. We're sitting in a restaurant having dinner. (I'll call you "Derek" for the moment). Here's a piece of our script for scene 24:

MIKE
Nice food.

DEREK
You gonna eat those egg rolls?

MIKE
No. You can have 'em. I'm stuffed.
(beat)
Say, what time is it?

DEREK
(looking at watch) 7:21.

MIKE
Shouldn't Jennifer have called by now?

DEREK
Uh oh, maybe something's happened.
Let's get out of here.

(They nod, and bolt)

This is the scene. Simple enough? What do you think it would take to shoot it? If you're thinking 20 seconds, think again. Even though the scene will cut to about 20 seconds (and appear to take 20 seconds of real time), it might take all day to shoot. Why? Unlike those sitcoms on TV where you see four big cameras rolling around shooting the scene, most scenes in a film are shot with only one camera. Realize that these cameras cost a fortune, and the company shooting the movie is renting them. So you usually need to get by with one camera.

Next, unless you are going to play the entire scene in a wide shot, which you could certainly do if you wanted it to look like you were in the audience of a stage play, you are probably going to want to cover the action from a few different viewpoints. You may want a close up on Mike when he says "Shouldn't Jennifer have called by now" because you want to see the concern in his eyes. You might want a close up on Derek when he says "Uh oh" and a close shot of a watch when he looks at it. To build suspense, you may want a viewpoint that is far away, with our heroes tiny in the frame, silently playing with their food before you punch in on Mike's close up once he realizes that Jennifer is missing. The point is that directors don't shoot a scene for you to hear the characters simply reciting dialog. They often want to use the photography to build emotion. This is done by shooting from certain angles, by using close ups and long shots, by using different lighting conditions, by moving the camera or not moving the camera, and so on.

So, to *really* shoot the scene, you may begin by placing the camera at the table, where Derek should be sitting, and shooting the entire thing with the camera up close on Mike's face. Derek doesn't even need to be there. He could be off getting lunch. Each time you move the camera to a new spot, it is called a **set-up**. And when you slate each set-up, it would be given a unique scene number. The overall scene number for this dinner sequence is "Scene 24". So the first set-up would be called 24A. Now, once you set up the camera, and the lights, and have Mike sitting there, you could shoot him doing the dialog once and it would use about 20 seconds of film.

Done? Are you kidding? That was just Take 1. Mike performed the scene once. But the director may want more emotion, and so they do it again. Take 2. They may shoot scene 24A three, five, a dozen times until the director is satisfied she "caught it". The director may not use every take she shot, but will circle a few good performances on a sheet of paper and "print" those circled takes. Now, the director moves the camera and the whole thing goes again, this time with scene 24B: the close up on Derek. This could go on for hours. Then the camera may be moved again, this time for a long shot. Both actors will be required in this shot. 24C. And so on. The over the shoulder shot: 24D. The reverse angle over Mike's shoulder back to Derek: 24E. The birds eye view looking straight down at the table: 24F. The Steadicam shot that starts on their last line and follows them out the door: 24G. The aerial photography: 24H. And that's just the shots with the director and the actors. It is possible that these people are all too important to hang out all day to do this. So at some point, another group of people might shoot the outside of the building at night, the close up of the watch, and, of course, the eggrolls. This alternate group is the 2nd Unit (they do 2nd Unit Photography). And all of this: a total of 45 minutes of film, must be edited together into the simple 20 second sequence you see on the screen.

The film goes to the lab to get developed (into negative) and then printed (into positive) and then synced with the sound. It's not altogether unlike dropping your vacation pictures off at the Photomat: you take in the film, and the next day you get prints and negative back. The editor will get all this raw material, and from it must construct the final product. But how?

When delivered, it is likely that the 8 scenes (24A through H), each with 3 or 4 chosen takes will be assembled together, end on end, for a long, repetitive, boring kind of show (called "dailies", because of the frequency of this ritual). The editor must watch this show, think about how the scene *should* look when done, perhaps talk with the director about her vision for the finished scene, and then go to work — pulling the best parts out of all the takes, and putting them together in a compelling and interesting way, all the while making the work invisible to the average viewer. The finished cut needs to flow *naturally*. There is no one way (and certainly no *right* way) to edit a sequence together. And remember, regardless of what was intended in the writer's mind, regardless of what the director was hoping to shoot, the only thing that really matters is what was captured on the film: for that is all the editor has to work with.

Another point worth noting: most films are not shot in order, that is, the order of the scenes as they appear in the script. On the first day of shooting, it is common to begin somewhere in the middle of the script (based on logistics and scheduling of actors, equipment, locations, and so on). This means that editors traditionally cut scenes in *shooting* order and not *scene* order. It's just another of the many joys of being an editor.

Although this handbook is not specifically about the editing of film or video, it is likely that many individuals working in videotape are unfamiliar with film, and vice versa. This section presents a brief overview of these two worlds of editing. Not all film or video projects will be exactly as described here, but it should provide you with a general understanding of the concepts and language involved.

THE FILM STORY

Francis Ford Coppola once described film editing as cutting up a dictionary and writing a novel with the pieces. At a mechanical level, editing is quite simple — cutting and taping.

Let's begin by examining the flow chart on page 26. Film is shot on *location*, along with production sound which is usually recorded on 1/4" audio tape or maybe DAT tape. The negative film you put in your Nikon 35mm camera is not all that different from the film loaded into a motion picture (like a Panavision) camera: it is 35mm wide, the film stock comes in small cans, usually from Kodak, Agfa, or Fuji.

When you get your snapshots back from the Photomat, you know how the film has those little numbers on the edges? 1 ... 1A ... 2 ... 2A ... and so on through your roll. These are called "latent edge numbers."

On motion picture film, these *edge numbers* (also known as *key numbers*), are divided into two parts: the *prefix*, which indicates what roll of film was used; and a *footage*, which increments once for each foot of film — usually 16 frames. These key numbers are extremely important because every frame of film must be uniquely labeled. Reference to a key number is the only way a specific frame of negative can be located.

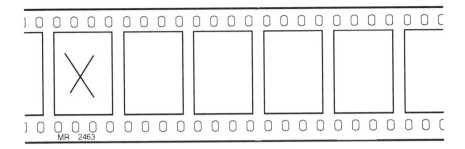

There is also a manufacturer's code on the edge of the film. This tells people what kind of film stock was used, what company manufactured the film, and so on.

Another difference between your snapshot film and movie film is that your roll is about 3 or 4 feet long, giving you room for 24 to 36 exposures (frames). Movie film rolls are longer, usually 1000 feet each.

The shape of the frame on a roll of film is NOT determined by the film

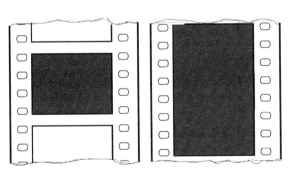

stock, but rather by the CAMERA: its mechanics and optics. The familiar shape and size of snapshot frames is different than the shape of 35mm frames shot in motion picture cameras.

Different kinds of motion picture cameras can even place the image on the frame in different positions and in different proportions. Panavision and Vistavision offer just two of the different ways frames can be placed on 35mm film.

Aspect ratio is the length-to-width proportion of a film frame, regardless of the actual size of the frame. Clearly, two frames can have the same aspect

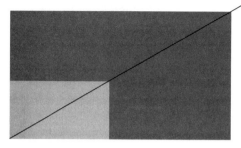

ratio but the area of one can be vastly different from the other. More area means higher resolution (or higher quality) of image. These two rectangles have equal aspect ratios but the larger one has more than four times the area.

The *sprocket holes* along the side of the film allow for toothed spools to move the film along in cameras (as well as in all film equipment). These sprocket holes are also called *perforations*.

For every frame on traditional 35mm film, there are four perforations along the side (hence the expression "4-perf film"). If you want to make the images a little smaller, you can modify your camera to shoot frames that only have three perforations

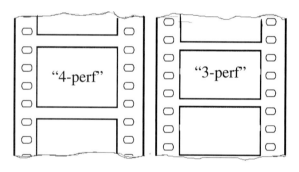

along the side, called "3-perf." This makes the same 1000-foot roll of film hold 25% more frames, and although the aspect ratio is changed and film area is smaller, it often provides acceptable image quality at less cost.

When a film roll is finished it is sent to the lab for processing, just like your own 35mm rolls. You are aware how expensive 36 exposures can be to develop and print; imagine millions and millions of exposures! Film productions would rather not waste money on prints of useless shots, consequently they develop every roll but only print selected scenes.

While each roll is being shot on a location, someone (often a Camera Assistant) takes notes as to what scenes are on each roll, how many takes were shot of each scene, and how much footage was used on each take. When a shot is completed, the director will often ask that it be "selected" or "circled," meaning that he liked it or might like it, and wants it to be printed. The Camera Assistant actually circles the take number on this "camera report" and this report goes with the roll of film to the lab.

Also during the shoot, the Script Supervisor takes notes on a copy of the script. These *continuity* notes describe each camera set-up, highlight circled takes, and "line" the script with vertical lines depicting each take. At the end of each day, these continuity notes along with copies of the camera reports are sent to the editorial department.

At the end of each day's shoot, the film and camera reports are sent to the lab for processing. The lab develops all the negative and then makes a positive

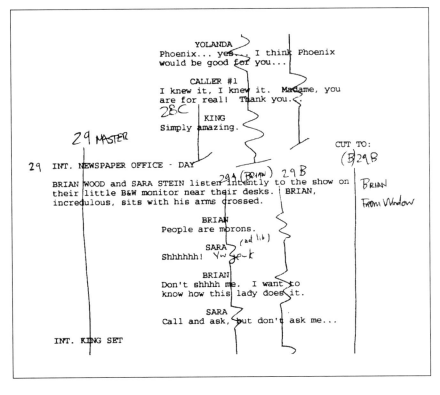

print of the *circled takes*. The negative that was not printed is considered no good (or N.G.) and is placed in a vault as "B-neg"; it will probably never be used but could be printed if later needed. The negative that was printed is vaulted also, but kept separately from the B-neg. No one wants to handle the negative very much. You don't want to risk ruining it with scratches or dirt or anything. Remember, after you have shot a 50-million-dollar movie, after everybody goes home, all you really have to show for it is the negative. It is delicate. It is irreplaceable.

Meanwhile, the 1/4" audio tape that was recorded on location (probably on a Nagra brand tape recorder) is transferred to another magnetic stock, this one with sprockets on the sides so that it is in the same format as the picture. This stock is called "sprocketed mag."

In the editing department, the print from the shoot (usually called *workprint*) and the sprocketed mag (now referred to as *track)* are delivered usually the next day. First, one of the editorial assistants will *sync* the picture and sound. This involves finding each scene's slate, locating and marking the "clap frame" and then doing the same on the mag. Once the clap point has been found in both picture and sound, the two strips of sprocketed film are lined up, marked, labeled, evened out, and built into larger rolls of dailies.

Then the film dailies are *coded*. Coding workprint and track provides a numeric reference to locate which track goes with which picture. Workprint and track are fed into a table-top coding machine (Acmade is a popular brand) that applies *code numbers* (also sometimes called "rubber numbers") to the celluloid, one number per foot.

Now the workprint has two sets of numbers on it: the original key numbers that came on the negative and can be read on the positive print, and the new code numbers that allow the editor to sync picture and sound, and identify every frame on the roll.

The code numbers and key numbers have no relationship to each other in value or position; they are completely different sets of numbers. So assistants build extensive *log books* that refer key numbers to code numbers, and code numbers to scene numbers and they make labels for every scene to go with the rolls of film.

For each day that film is shot, it is sent to the lab, developed and printed overnight, and returned to the production the next day.

When there is time, the editor, director, cinematographer, and anyone else of sufficient importance gather to watch dailies, (or rushes) in a screening room. They screen all the previous day's film, perhaps discuss shots the director likes or how he'd suggest approaching the edit, and then other acting or technical matters are evaluated.

Finally, the editing can begin.

MOVIOLA

The dailies are "broken down" into small rolls, each containing one take. Racks of these little rolls are brought to the editor, who begins watching the film on a machine called an *upright*. The most widely used upright editing machine is manufactured by Moviola. The upright "Moviola" was invented in 1924, and has remained virtually unchanged since then.

Film is fed into the Moviola; it can be stopped and played, forwarded and reversed, at play and slow speeds. The editor uses foot pedals to control the machine. With a grease pencil, also known as a *china marker* (marks made are sometimes called "chinagraph marks") the editor marks frames, then pulls the film from the machine, and with a splicing block to hold the film, cuts it between frames.

Film dailies are cut apart and taped into sequences. The leftover portions of the dailies, the *head and tail trims*, are hung up on small hooks and draped into a large cloth basket — known as a "trim bin." Each piece of film must be labeled so that it can be later found if needed. Editors might have many bins surrounding them while working.

Many editors work with two Moviolas: one for watching and cutting dailies,

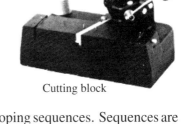

Cutting block

and one for playing and re-cutting the developing sequences. Sequences are strung together until their total length is about 1,000 feet — these are called "reels." A thousand-foot reel is about the largest manageable load of film, although reels can be built larger, up to 2,000 feet, if necessary.

An average feature film starts with about 150,000 feet of selected dailies (equivalent to about 27 hours of film), and is edited down over the course of about 12 weeks to just under 2 hours (about 12 reels of film). Some movies begin with 500,000 feet or even a million feet on rare occasions.

A typical film room might also have a flatbed editing table for editing or screening. The most popular manufacturers of flatbeds are KEM and Steenbeck.

Some editors do not work on Moviolas, but prefer the flatbed approach. Other editors find the flatbeds cumbersome.

Either way, after months of film editing, a final cut is chosen, and the much-re-spliced workprint is sent to the *negative cutter*, along with a copy of the log book. Using the key numbers to locate negative, the negative cutter carefully cuts and splices the original film negative, using a hot-splicer to virtually melt the ends of the shots together. The workprint is used as a guide to cutting the negative, to insure total accuracy. Negative cutting for a film takes just over a week.

courtesy of KEM

A KEM flatbed

The final negative is later matched back with the final mixed tracks of audio (which, when completed, is printed as a single *optical track negative*) and dubbed to an "answer print" positive to check and modify color balance between shots. Answer prints may sometimes go through a few *trials* until the color timer gets all the colors right, and everyone likes how the film looks.

When these choices are finalized, an *interpositive print* or IP is created. This is a positive print of the film on negative stock, and is basically a protection copy for the original negative. From the IP, *dupe negatives* are produced, usually a couple, from which all *release prints* of a film will be created. The very first release print is also used as a *check print*, to verify the color timings after all the generations of dubs have been completed. If colors are still okay, all the remainder release prints are copied and sent to theaters around the country.

If this movie were for television, the final process would be slightly different. Rather than make all the expensive prints and dubs, the original negative of the **cut** film would be sent to a videotape post-production facility and *telecined* (transferred) to broadcast quality videotape, usually 1". Until electronic post production for television began changing the broadcast industry in the mid-80s — by allowing for source negative to be telecined, edited on videotape, and delivered on both film and video — this cut-negative telecine was the primary way film-originated material was shown on television.

THE VIDEO STORY

Videotape projects often start with a film shoot. If not originating on Betacam or 1" videotape, the project is often shot on 35mm or 16mm film. These film projects will finish on tape; so although they go through the 3:2 pulldown in telecine, they do not track film information. The telecine process gets the film into the video domain, where it will remain indefinitely.

Telecine records the film onto 1" (or Digital Betacam, or D2, D3, D5, or DCT) raw stock, with a 3/4" tape recorded simultaneously during the session (sometimes referred to as a "simo"). This 3/4" tape is recorded with the same timecode as the master 1" tape, and a visible timecode window is generally "burned" into the lower portion of the frame.

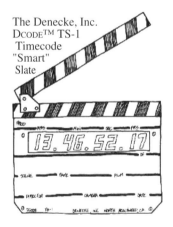

The Denecke, Inc. DCODE™ TS-1 Timecode "Smart" Slate

The audio, recorded on location using a Nagra 1/4" audio recorder (or increasingly on a Digital Audio Tape "DAT" recorders), is synced up by the telecine operator (known as a "colorist"). To facilitate the somewhat slow process of syncing the 1/4" audio and the negative, many productions utilize special *Smart Slates*. These are clap sticks that have a timecode

displayembedded in them. Timecode is fed to both the Smart Slate and audio tape and the colorist can use these numbers to quickly and electronically sync the dailies in telecine.

For music videos, or projects with playback audio, the 1/4" music tape is synced to the source

Courtesy of Rank Cintel

A Rank Cintel telecine.

picture, and laid down on the tape. The 1/4"'s playback timecode is also burned into a separate window on the simultaneous 3/4" tape and occasionally placed into the user-bits of the videotape (in telecine if not during the production shoot).

Telecine records tapes until they have no more than an hour of source material on them, and often less. Once completed, the 1" master tapes are stored in a special vault by the video post-production facility for use later.

The 3/4" tapes are taken to an offline bay and screened. Often, during screening, a frame-grab printer is used to take "snapshots" of the tape at desired frames. This machine will print copies of selected video frames onto a special roll of paper ("thermal paper," like that used in FAX machines). Since the 3/4" tape has a visible window containing the timecode, rolls of these images can be built that act as a select log of sorts.

Using a keyboard, the editor types timecodes for cueing, or can mark edit locations "on the fly" (while the tape is playing), simply by pressing "mark in" or "mark out" buttons at the appropriate moment. Using the keyboard edit controller, the editor can mark ins and outs of shots, preview them, and record them onto a separate 3/4" tape called the "record." Edit controllers are manufactured by CMX and the Grass Valley Group, as well as by other manufacturers, all of which offer a range of systems that vary in price and functionality. Typical offline controllers include the CMX 3400, the GVG 141 or 151, the Sony BVE 910.

A CMX 3400 edit controller.

Linear edit systems consist of a keyboard controller, display monitors, and some kind of central computer interfaced to necessary machine controllers — running the peripheral tape transports.

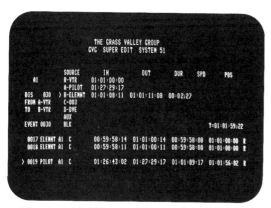

The System 51 editing screen from the Grass Valley Group.

Since it takes time for a videotape deck to move a tape from a dead stop to 30fps, you need a kind of running head start before watching a tape. *Pre-roll* is the distance a tape is backed up before the in-point of an edit in order to be moving at the correct speed at the edit point.

For a shot to be recorded properly, both the record tape and the source tape have to be moving together at the right speed. Consequently, editing systems must calculate an interlock relationship from the appropriate edit point, then pre-roll both machines a certain distance back, still synchronized up, then roll forward until they are at speed. Once at speed, the recording will begin at the chosen in-point. If one or both machines aren't moving at the correct speed, or fail to reach the edit point in time, the interlock relationship might be lost, and the edit will be aborted automatically. An editor wants sufficient pre-roll time, without making it so long that time is being wasted. Five seconds is usually enough.

Slowly, linearly, an edit decision list (EDL) is built. This list is a chronological history of the editorial decisions made. As each edit is recorded, an entry is placed in the computer memory. If a shot is modified, often both the old and new versions of that event remain in the list.

Monitors in an edit bay generally present the "record" tape large, centrally, and in color. This monitor will either double for "previews," or there will be a separate monitor adjacent to it. Secondary source monitors are located peripherally. These smaller black and white displays are provided for each source videotape deck. This way an editor can see where all the tapes are located.

There is often an *effects switcher* in the bay for doing simple effects and keys (superimposing images). The most commonly used switchers are the

Grass Valley 100-series. To create dissolves in offline, the outgoing and incoming shots must be on different tapes. If both shots are on the same reel of source material, as they often are, a *B-roll* must be created of the desired shot. This reel, named for the original source tape but usually with an appended "B", will be fed into the effects switcher along with the origi-

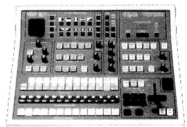

A VGV DX-120 effects switcher.

nal source reel, and the switcher will superimpose the two as the effect dictates. Because of the technical properties of video, it is important that both images be synchronized not just in time but in framing and color. A peripheral device called a *time base corrector* (TBC) is used to balance both horizontal and vertical signals against visual timing anomalies like "jitter." TBCs are also used to synchronize multiple video decks' signals. In general, visual effects cannot be done in analog without a TBC on each videotape machine.

A "key" is a special superimposition of two or more video images. Keying is a complicated area of video editing. To simplify: one image, called the "matte," is a video element with holes in it through which other images can be seen. Sometimes, the key does not involve a matte with "cut" holes in it, but rather a background of one of the primary colors (red, green, or blue) that can be replaced by a keyed image. This *chroma key* is an older type of video keying. It is commonly seen in news and weather reports.

The offline editor edits and re-edits certain events, working in somewhat random ways, and will create numerous entries in the *edit decision list* (EDL). Some of these events will be current, many will be redundant, new shots will be added, and others removed; the list is not usually orderly nor is it easy to follow. Re-organizing the events of an EDL can be laborious and difficult; consequently, most EDLs are "dirty."

Once the offline session is completed, the "dirty" EDL should be "cleaned" in order to maximize the efficiency of the (expensive) online session. There are programs that will clean EDLs — the original for this was ISC's "409." Another significant program that cleans up EDLs is called "Trace," also originally developed by ISC. With Trace, the edited master tape can be moved from the record deck to a source deck, if desired. Now, with the edited tape as a source, new edits can be made from it to yet another record tape. In an effort to cull material from larger numbers of source tapes to fewer ones, this process is common in video editing. However the EDL will not automatically be able to follow where a shot originated after it has been copied from source to master a number of times. "Trace," using reel identification numbers and timecodes, will compile a single EDL that can re-create the edited master from the *ORIGINAL* locations of each source shot.

The EDL is copied from the editing system onto a floppy disk, formerly a 5 ¼" size but 3 ½" floppies are now generally the rule. Whatever the size, this EDL must not only be in the proper format (CMX, GVG, Sony, *etc.*) but the disk format itself must be appropriate for the online system (Mac, DOS, RT-11, *etc.*). To-gether with a printout (also called a "hard copy") of the EDL, the infor-mation is brought to online for re-creation using the master elements. Where offline might cost anywhere from very little to a hundred dollars an hour, online begins at a couple hundred per hour, but with digital equip-ment and full effects, it might cost $1,000+ per hour.

Once the master tape is final-ized (and the EDL is *locked*), before, during, or after the online, the audio might be "sweetened" in a multitrack audio session. *Sweetening* is the post-production of audio for videotape masters (also called "mix-to-pix"). Traditionally, production tracks are laid down onto one or two of perhaps 24 tracks of a 2" audio tape machine. Although not quite as flexible, the 24 tracks of an audiotape are like 24 separate dubbers in a film sound department. Using SMPTE timecode to correlate sync between picture and sound, ADR, effects, and music can be edited to the 2" tape for mixing and eventual layback to the master videotape. In the mid-80s new nonlinear audio technology was introduced, called digital audio worksta-tions (DAWs) that subsumed these multitrack mixes in the digital domain.

A BRIEF HISTORY
OF ELECTRONIC EDITING

Film editing has hardly changed since the upright Moviola was introduced in the 20s. Even after the flatbed was developed, although film editors often chose one "system" or the other, both involve many mechanical similarities and identical media.

Through the 40s and 50s, television was born and grew. TV was a strange and arguably inferior medium than the theatrical presentation of film, but it was a growing trend and unmistakably pervasive. In its early years, programming was LIVE, broadcast directly from studios in New York to viewers across the country. It was not edited, and thus didn't involve the editing community — it was just this other "thing."

But since Los Angeles time is three hours earlier than New York time, and LIVE television is seen *simultaneously* all over the country, there was no way for the early networks to provide ideal scheduling of shows. They wanted a way to delay West Coast broadcasts. But how do you delay a *LIVE* broadcast?

1956

◆ The delay problem is solved. A company in California — Ampex— invents an electronic recorder for broadcast television: the first *videotape recorder* (VTR) controls a 2" roll of magnetic tape and is called the *VR-1000*.

To properly set the stage, we must first recognize that of the three networks, CBS was clearly the leader. Not only did they have the top-rated shows, but also had a powerful research division called CBS Labs. Although a branch of NBC was actively doing research in mechanisms for using the new videotape, CBS was in the unique position of having both film and video (electronic) production companies in CBS Television: an electronic facility in New York and both electronic and film facilities in Los Angeles. At the LA film studio (in Studio City), they shot shows like "Gunsmoke." CBS Television was in a position to know both worlds well and recognize the advantages and drawbacks of each.

The director of engineering at CBS was Joe Flaherty, a visionary in broadcast electronics and considered by many the greatest single force in the advancement of broadcast editing technology. If Flaherty had a mission, it was to move television production from film to videotape. Film shoots tend to shoot too many takes to cover themselves; you can never be quite sure that you got the shot you wanted. Flaherty believed that with tape, by seeing the material immediately, you could save shoot time, location days, actors' time, *etc.* . . . in other words, save money.

1957

✦ All three networks purchase the new videotape recorders and begin experimenting with ways to edit the videotape.

Originally, editing videotape was like editing film or audiotape — with a razor blade and adhesive tape. Unfortunately, unlike audio, video must be cut between frames, and there is no visual guide as to where a frame boundary is located. Editors used a chemical mixture containing extremely fine iron powder that, when applied to videotape, made the 1/200-inch space between frames — called the "guard band" — somewhat visible. A "clean" edit between frames was difficult; cut transitions frequently lost picture signal and broke up. Videotape editing was extremely hit or miss. In the following years, an expensive (over $1,200) cutting block was developed, called a "Smith splicer," which allowed the editor to see the guard band through a microscope. Still, there was a differential between where audio and video events occurred on the tape; even sync-cuts involved careful and slow measurements of picture and audio.

The "Smith Splicer",
1957

Engineers at NBC developed their own in-house methods for editing videotape that they called the "edit sync guide" or ESG. This was a track with an electronic beep every second, followed by a male or female voice announcing each minute and second. Coding the videotape and a kinescope (film copy) of the videotape with identical ESG codes enabled television to be edited via traditional film methods. Later, the ESG was used to help physically conform the master videotape, much in the way negative cutters use workprint to conform negative. NBC's methods were so successful that other networks often brought projects to them for editing.

It is somewhat ironic that in the late 50s, video editing was using film as its working media; by the 80s, film editing was beginning to use video.

1962

✦ The first commercially available electronic editing system, the *Editec*, is developed by Ampex. ✦ Time clocks are recorded onto magnetic tape; applications in editing are investigated.

The Editec involved the selective recording of one videotape (source) onto a target videotape (master). Since there was still no visible way to tell where you were on your tape, electronic pulse tones were recorded along the length of the tapes at desired edit points. An electronic edit meant a cleaner splice and the audio/video differential in cut point was rendered moot. But previewing or exactly re-creating an edit was impossible; you simply had to try it again. The problems were 1) the frame accuracy of the point the edit started re-

The Ampex Editec, 1958

cording on the master tape, and 2) how to synchronize the source tape and master tapes while rolling. These problems were solved very slowly over the following years.

In the mid-1960s, Dick Hill, a CBS technician in Los Angeles, happened upon a piece of Defense Department technology that allowed electronic recording of a time clock on magnetic tape. When the military was testing missiles, they had devices monitoring from various locations around the test site: some at the launch point, some thousands of miles away. Launch and missile data were recorded on magnetic tape, and along with that information was a time-of-day clock, which allowed them to track and relate events occurring at different data recording locations.

Dick Hill felt this technology might solve the pulse tone problems in editing and began sending reports to CBS headquarters in New York. Adrian Ettlinger, a CBS engineer, was sent to investigate Hill's reports. Ettlinger reported back to CBS that this was a good thing, and that they should encourage the development of what was to become known as *TIMECODE*.

1 9 6 7

✦ EECO (The Electronic Engineering Company), in Orange County, California, begins manufacturing the first timecode equipment.

With timecode available, CBS Labs began developing an editing system to utilize these numbers. Ettlinger, with other engineers, first experimented with a system of Sony recorders that could perform a continuous play of an edited sequence by accessing a number of duplicated source tapes — the origin of "look-ahead" previews. But the 2" tape machines were unwieldy and too many would have been required to achieve any useful kind of nonlinearity. Later they began using computers to control various kinds of newly-developed 1/2" videotape machines, with mixed degrees of success.

1 9 6 8 - 6 9

✦ NBC's "Laugh In" pioneers the use of videotape editing as something more than an extension of live switching. The show has 400 - 450 physical edits where most other shows have between 40 -100.

By 1969, none of CBS Labs' tape format experiments had proven particularly appealing, so they abandoned tape and began experimenting with magnetic disk platters for storing the analog video information. They ran into some mechanical problems involving the disk platters and turned to the Memorex Corporation for assistance. Memorex was intrigued with exploring the technical feasibility of recording video on disk media for random access.

1 9 7 0

✦ In January, CBS Labs and Memorex create a joint venture called CMX to make editing systems. They build the CMX 200 and CMX 600.

The business plan for the CMX (which stood for **C**BS, **M**emorex, e**X**perimental) venture was drawn up by Adrian Ettlinger and Bill Connelly (both from CBS), and Bill Butler and Ken Taylor (both from Memorex). Butler became the company's general manager. Their first product had two parts. The first part, the *CMX 600*, was a computer with a stack of removable disk platters, each holding about 5 minutes. The platters looked like a horizontal bread slicer — their cost: $30,000. With six disk drives working in union, the 600 could locate any frame in under a second from about 30 minutes of mediocre (half-resolution) black-and-white dailies. This part was a nonlinear editor, using what would become SMPTE timecode. It did not produce a master videotape; rather, a punched paper tape encoded with

The CMX 600, as shown in the original CMX product brochure, 1971

a list of timecodes for re-creating the edits on a broadcast-quality machine. The second part of the system was the *CMX 200*, a linear tape assembler that would take as input the 600's list and build a master 2" broadcast videotape.

1971

✦ The CMX 600/200 combination costs $500,000 and is patented by CBS; its inventor is Adrian Ettlinger. In April it is demonstrated at the CBS stockholders' meeting.

The first commerical application of a nonlinear editing system, the CMX 600, is begun. CBS Television used the the system on a movie-for-television called *Sand Castles*.

1972

✦ After a number of years of investigation, the Society of Motion Picture and Television Engineers (SMPTE) adapts the EECO timecodes and standardizes their industry use.

By this time, five CMX 600s had been built — besides placements at CBS and Teletronics in New York, one was located at Consolidated Film Industries (CFI) in Los Angeles. Head editor there, Arthur Schneider, formerly with NBC, and Dick Hill (formerly at EECO) worked with CMX programmers, among them Jim Adams and Dave Bargen, to develop and beta test improvements to the systems. Problems in the CMX 200 led to the CMX 300. Where the 200 could only take input from a 600's printed papertape, the 300 allowed for a keyboard terminal to interface directly with the assembly editor. CFI suggested that there was a wonderful product in the linear online system, the 300. Both for practical reasons (perhaps due to its high cost, poor image quality, high maintenance requirement, and limited storage of source) and for social ones (it radically changed all accepted work rules and production concepts), the 600 did not catch on; however the linear editing products that CMX developed did.

At the National Association of Broadcasters (NAB) show, the CMX 300 was demonstrated. With guidance from users like CFI, the CMX 300 continued to evolve; by 1978 it had become the 340.

1973
✦ Sony introduces 3/4" tape. ✦ The CMX 50 is debuted at NAB.

Sony introduced the first 3/4" machine (and a standard for 3/4" tape called U-matic) with hopes of seeing it adopted for home television recording; but it was too expensive and too large for consumers, and ultimately did not take off. On the other hand, because it was more cost effective than previous editing formats, 3/4" videotape was picked up by the broadcast and industrial markets.

At the NAB show CMX introduced the CMX 50, the first commercially available 3/4" edit controller — because of the inexpensive U-matic tape machines, it was considered the first viable "offline" system. The product used a standardized Edit Decision List (EDL), look-ahead previews and assemblies, and a style of editing that continues today. With the commercial

The keyboard controller for the original CMX 50, 1973

success of the 300 and the 50, CMX wanted to expand; however, Memorex was financially strained, and CBS would not contribute more than 50% of company ownership. The decision was made to sell the upwardly mobile company.

1974

✦ Orrox purchases CMX and names Bill Orr as president. CMX's manager of product development, Dave Bargen, leaves the company. ✦ After two years of development, two separate corporations — MCA and N.V. Philips — announce that they are working together to standardize formats for their new invention, videodiscs.

1975-76

✦ The MCA/Philips agreement produces a format for optical video discs known as the LaserVision standard. Around the same time, a number of companies, among them Sony, Philips, and Hitachi, announce their research into laser optical audio discs.

Due to popularity of the systems among existing CMX clients like CFI and Vidtronics, Dave Bargen began writing software programs that would increase the functionality of the CMX systems. First he developed "409" (an EDL cleaning program), then "Trace" (another special list management tool), and finally "Wizard," (which became "Super Edit"). Eventually, Bargen marketed these products to other CMX clients.

1977-78

✦ The CBS-Sony system is created. ✦ Laserdisc players are first sold. ✦ A film from young director George Lucas dominates at the box office. *Star Wars* begins what will become the largest ticket sales in the history of movies; the trilogy of films will gross over 4 billion dollars in ticket, cassette, and ancillary product rights over the next ten years.

Because Joe Flaherty still believed that television should be done on videotape, CBS Labs continued to develop systems. Late in 1977, CBS began an advanced form of their CMX 600 project, this time using a new type of 1/2" videotape that was being developed by Sony called "Beta." This new format unfortunately required expensive tape decks. CBS's new editing system used modern computers and the interface, like the 600's, was through a lightpen. Adrian Ettlinger was moved from consulting peripherally on the project to being the system's software product manager. By 1978 the CBS-Sony system, as it was called, was in use at CBS. But both companies decided not to pursue the marketing and manufacturing of the product, and it remained in-house at CBS.

In 1977, the first LaserVision videodisc players were sold in the educational market. MCA teamed up with Pioneer to form a venture, Universal Pioneer Corporation (UPC) to mass-produce videodisc players. Magnovox introduced their competing disc format, "MagnoVision," utilizing completely different technologies to play video discs.

1978-79

✦ New video formats deluge the market. ✦ ISC is formed. ✦ Philips demonstrates the first audio "compact" discs in May. ✦ Bell and Howell acquires Telemation. ✦ CMX invests in DBS; begins developing new kinds of editing systems. ✦ Lucas and Coppola investigate video applications to filmmaking.

A host of new video formats began to inundate the consumer market: Sony's *Beta* format on 1/2" videotape; Panasonic also had a 1/2" tape format called *VHS*; Magnovox's *MagnoVision* videodiscs; MCA and partner IBM's 5-year-old entity DiscoVision Associates (DVA)'s 12" videodisc format, *LaserVision*.

Dave Bargen, formerly of CMX, worked with a former chief engineer of Vidtronics, Jack Calaway, who had built a somewhat more flexible machine controller interface than the previously best-known CMX I^2. Bargen used a DEC computer, Calaway's hardware, and his own CMX-like software to create a new editing system. He formed the Interactive Systems Company (ISC).

In mid-1979, CMX/Orrox's penetration into the equipment market was beginning to plateau (over 90% of all broadcast editing in 1978 was on CMX equipment). CMX/Orrox saw the Direct Broadcast Satellite (DBS) business as the next boom industry. In a bold move, they began to invest heavily in DBS.

At the same time, development began on the newest CMX products, the 3400 and 3400 Plus. The plan was to move videotape editors smoothly from somewhat difficult and number-intensive editing (in the 340) to a modern editing system: the 3400 Plus would have a database management system (DBMS) and soft-function keys, and would implement a new computer feature, a windowed graphical interface.

Also in 1979, George Lucas began investigating ways of improving the filmmaking process. His friend and mentor Francis Ford Coppola had been active in using video technologies to help in production. Coppola and Lucas, like their friend at CBS Joe Flaherty, understood the cost savings that could be achieved on location by shooting in video instead of film — by being able to view immediately the material you had shot. Since neither felt video looked as good as film for production, it was generally understood that the video would only be a tool in the film process.

Coppola had pioneered the use of a video camera alongside the film camera on shoots — the video was fed to and recorded in a customized trailer that sat at his locations, called "the silver bullet." With his video specialist Clark Higgins, Coppola is considered the first to use *video assist* in film productions, in particular on 1982's *One From the Heart.*

Lucas was more interested in facilitating post production by using computers and video technology, and hired an expert from the New York Institute of Technology (NYIT), a leader in academic computer applications, particularly in computer graphics and animation. Ed Catmull moved to Lucasfilm in California where he began investigating the post production process and planning what type of technologies could be used.

Bell and Howell, a Chicago-based film equipment company established more than a half-century earlier as a manufacturer of 35mm film printers, purchased an equipment distributor/manufacturer called Telemation. Bell and Howell's Video division (consisting mostly of tape duplication) hired Jim Adams for Telemation, and in doing so, acquired his previously developed editing system, the *Mach-1.*

1 9 8 0 - 8 1

✦ German equipment manufacturer Bosch-Fernseh purchases 50% of Telemation from Bell and Howell in 1980; by 1981, it acquires the remaining 50% (and the *Mach-1*) from Bell and Howell, and calls the new company Fernseh, Inc. ✦ Early in 1980, George Lucas creates his Computer Division. Late in 1981, Lucas releases his third blockbuster, *Raiders of the Lost Ark,* directed by friend Steven Spielberg. ✦ Adrian Ettlinger begins work on his ED-80 editing system.

At Lucasfilm, Ed Catmull presented a plan for the development of three products: a picture editing tool, a sound editing tool, and a high-resolution graphics workstation. The proposed development cost of this three-project plan was about 10 million dollars. It got the go-ahead after the success of Lucas' new film, *The Empire Strikes Back.*

By the end of 1980, Catmull hired three computer experts: the picture-editor (which became the *EditDroid*) was led by NYIT friend Ralph Guggenheim — also an NYU film school graduate; the sound-editor (which became the *SoundDroid*) was led by Andy Moorer — a digital audio pioneer who had been with CCRMA (the Center for Computer Research in Music and Acoustics at Stanford University); the graphics-project (which became the *Pixar*) was led by Alvy Ray Smith — a graphics expert from NASA's

Jet Propulsion Lab (JPL) and before that, Xerox PARC (their research division). By late 1981 the Lucasfilm Computer Division was actively developing these products.

But on what kind of workstation would they be run? At the time, there was no such industry as "desktop computers," and a number of companies were competing against the giant IBM for the market of smaller high-powered computer workstations. The host computer had to be relatively small, fairly inexpensive, and of high enough horsepower to run simultaneously all the things these systems had to do. Lucasfilm investigated all options, tried a few different hosts, and finally chose a SUN computer (with Motorola's 68000 processor) costing around $25,000.

As for what video media would be the source for the EditDroid, the decision was made in Catmull's '80 report. At the time there were only a handful of potential formats: 1/2" Beta, 1/2" VHS, LaserVision and MagnoVision 12" videodiscs. Catmull, and later Guggenheim, saw laser-discs as the defining characteristic of the EditDroid project — they were convinced the price of laserdiscs would drop, that a consumer adoption of discs was just around the corner, and that discs were the only way to achieve any kind of real nonlinearity.

Lucasfilm began talks with leaders at DiscoVision Associates about the possibilities of recordable LaserVision discs. The EditDroid, at the time called the *EdDroid*, was a nonlinear, computer-controlled, disc-based editor. The system would control tape (because you had to control 3/4" for *emergency* linear editing), however the thrust of the development was on a new "style" of editing, manipulating the laserdiscs, and the idea that the human interface was going to be significantly more "user friendly" than computers or video editing systems.

The point-and-click computer style, mouse-based control, bit-mapped and icon-based graphics were unseen in the consumer world, and only beginning to take the academic computer world by storm. The Computer Division's products all utilized these features.

Throughout the early 1980s, numerous individuals and organizations came through Lucasfilm's Computer Division to see what was going on. Among them, an English entrepreneur named Ron Barker, who had been working outside of Boston, and who was interested in developing a revolutionary new editing system of his own. Also visiting the team was Adrian Ettlinger, who had much earlier left CBS Labs and had also been developing a new editing system of his own.

Ettlinger's system was based on a smaller computer, the Z-80, and was originally called the *ED-80*; named for the host computer and for the year in which his project began.

1982

✦ The Montage Computer Company is formed and work begins on the Montage Picture Processor®. ✦ The Optical Disc Corporation is formed.

Ron Barker teamed up with engineer Chester Schuler and formed a new company outside of Boston, The Montage Computer Company. They began development of a tape-based nonlinear editor, using the 1/2" Beta format videotapes as their source media. They called the system the Montage *Picture Processor®* because in many ways it was like a word processor.

DiscoVision Associates (DVA), the venture between MCA and IBM, was shut down and most parts were sold to former-partner Pioneer Electronics. DVA remained as a small company holding the rights to all the laserdisc-related patents. Members of the DVA development team joined forces to start a company to develop a recordable laserdisc, initially for the editing market. They called the company the Optical Disc Corporation (ODC).

Lucasfilm hired Robert Doris to take over the management of the Computer Division. A fourth project that had been in the works for some years under the Licensing Division was added to the Computer Division: Games. For the next 18 months there was a great deal of development activity both at Lucasfilm (outside of San Francisco) and at the Montage Computer Company (outside of Boston).

1983

✦ A prototype of the Ediflex is built. ✦ BHP spins off as an independent company. The TouchVision is announced. ✦ Lucasfilm and Convergence create a joint-venture for the Droid products.

During its time at Bell and Howell, the Mach 1 interested members of the R&D division, called Bell and Howell Professional (BHP). They were interested in potential applications of video technology for film. Bruce Rady, with the rest of the R&D team at BHP, began working on a new video-for-film editing system project which they called the *Envision*.

When Bell and Howell was taken over in a leveraged buyout, they began to analyze ways to divide up and sell off the company. Management chose to drop all things "video" related, and the Envision project was dismantled. The members of BHP were encouraged to purchase their division from the Bell and Howell parent company, which they did; Bruce Rady bought the rights to the Envision project for developing on his own time.

Rady began designing his own hardware and software, in particular an extremely low-cost VITC reader, for use on his system — its name was then changed to *TouchVision*.

CMX/Orrox decided to close its DBS manufacturing division. Changes in microprocessor technology along with other factors left them unable to recoup their huge investments. At the NAB show CMX/Orrox debuted their 3400 and 3400 Plus to crowds; following the show, CMX began layoffs and eventually stopped development of the 3400 Plus. Its designer, Rob Lay, moved to Lucasfilm's computer research department to work with the assembled team on the editing project. CMX's slow-down in the editing equipment market allowed other companies to enter the arena and offer new products. ISC began to erode CMX's high-end equipment line, and new smaller companies formed to move on lower-end systems.

Ettlinger, now with partner Bill Hoggan, completed a prototype for a new editing system with specific applications for the television market in Los Angeles. He worked with a number of editors and designed a script-based editing system that would control 1/2" VHS videotapes as sources. The system, no longer the ED-80, was called the *Ediflex*.

Lucasfilm's Computer Division, under Robert Doris, begin looking for an equipment manufacturer to license and sell the Droid products they were developing. Their investigation turned up seven companies, among them CMX, ISC, Grass Valley, and Convergence. Lucasfilm considered the ideal partner to be a privately-held company (like itself), one that was relatively profitable, and one that was ready to sustain a joint venture. In September '83, Lucasfilm Ltd. and the Convergence Corp. signed a joint-venture agreement for the Droid products.

1 9 8 4

✦ The Montage and the EditDroid debut at NAB. ✦ ISC merges with the Grass Valley Group. ✦ Apple introduces the Macintosh personal computer. ✦ ODC begins shipping laserdisc recording systems in the Fall. ✦ The Laser Edit facility opens in Hollywood.

In late '83 and into '84 both the Montage Computer Company and Lucasfilm had been teasing the market with advertising and promotions about their new film editing systems. For the EditDroid, Convergence was beginning to manufacture the necessary hardware, and Lucasfilm was continuing the development.

NAB 1984 was the debut of the two editing systems, to a great deal of interest. Although both presented prototypes, neither the EditDroid nor the Montage was market-ready. The EditDroid had to deal with the high costs of making laserdiscs and also the software difficulties associated with releasing a product running on the UNIX operating system. The Montage had to deal with the seventeen 1/2" consumer Beta machines that the editor controlled. Betamaxes had no provision for machine control, and so each machine on each system had to be specially fit with a custom machine-control card. (The cost of 17 *professional* videotape machines would have been prohibitive.)

By the summer, Lucasfilm and Convergence were still hammering out the nature of their joint agreement. Lucas decided to spin off all parts of the Computer Division; Convergence, still running its own independent business, signed the spin-off papers officially creating a new company with Lucasfilm. Negotiations continued for more than six months. Guggenheim, the leader of the edit-

The original product nameplate for the EditDroid, 1984

ing project moved to the graphics project, and Robert Lay was moved into his position.

ISC, after having grown tremendously since its inception, was merged with Grass Valley, an almost 20-year-old company known for its video production switchers, graphics and routing equipment.

Apple released the first desktop personal computer to use bit-mapped and icon-based graphics and a Motorola 68000 processor at under $5,000 — the *Macintosh*. Advances in memory and processor-power began to dramatically affect the "computer-consciousness" of the public, with powerful computers that were easy and affordable.

By spring, a new post production facility was formed in Burbank, called Laser Edit, Inc.; it owned its own in-house system called the Spectra-Ace (developed by Spectra Image exclusively for them). Laser Edit was the beta site for the prototype of an optical disc recorder made by the ODC, and developed over the previous two years. The Spectra system was similar to conventional *linear* videotape editing systems except that dailies were delivered on ODC laserdiscs and used in specially-developed two-headed laserdisc players. Although disc-based, the system was linear and proprietary at the Laser Edit facility. President Bill Breshears initially marketed the system to multicamera television productions.

1985

✦ Interscope purchases the Montage Computer Company. ✦ Cinedco is formed to manufacture the Ediflex. ✦ The CMX 6000 project begins at CMX. ✦ Film-originated television programming begins testing the new systems.

In February, the Montage Company sought a new finance partner; Interscope Communications (which had just purchased Panavision from Warner Communications) bought the company from founder Ron Barker and brought in new management. The company was again on solid ground.

At the NAB show both the *Montage* and the *EditDroid* were shown again, this time considerably more stable than as prototypes the previous year. A prototype for the *SoundDroid* was also displayed by The Droid Works (the new Lucasfilm and Convergence company). The cost of the Montage and the EditDroid were comparable, well over $200,000 each, and both already had placed a few "beta" systems in the field.

Also at the show was a prototype of Bruce Rady's film-style editing system, the *TouchVision*. As a developer from the film equipment business, Rady was relatively unfamiliar with the video editor's world. But in extensive interaction with film editors, Rady learned that 1) film editors didn't really want computerized equipment, and 2) every editor described what he or she did differently, so coming up with a grand new simplified model of editing would be problematic. Basing the TouchVision on a flatbed, his system essentially allowed editors independent control of the source decks, instead of disguising the multiple decks necessary for achieving nonlinearity. By being able to lock and unlock various decks, editors were allowed to work as if using an electronic (videotape) multi-plate flatbed. The TouchVision prototype controlled three 3/4" Sony decks for source and was target-priced around $100,000.

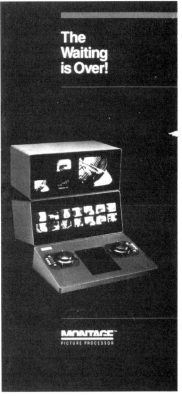

The Montage brochure cover, 1985

In May, Milton Forman and Adrian Ettlinger formed Cinedco to manufacture and rent Ettlinger's *Ediflex* system. Ediflex was not at the NAB show, because as part of a cautious marketing plan, it was only *rentable* in Los Angeles, for around $2,500 per week. By approaching the film market in a way it understood — system rental — the Ediflex, too, was being tested for use in the post-production of the 1985-86 episodic television season. The script-based system originally controlled eight JVC-400 VHS decks with its central computer, and used a light-pen to interact with the screen. Also for rental-only was Laser Edit's *Spectra System*, in Burbank.

CMX began to develop a new nonlinear system specifically for film editors. Seeing nonlinear as the direction all editing would go, CMX developer Robert Duffy became project leader on a completely disc-based system that will become the *CMX 6000*, a modern incarnation of CMX's earlier 600, and bearing little resemblance to their existing linear products.

Lorimar, more than any other studio, experimented with electronic post under Gary Chandler, Lorimar's vice president of post-production for both television and theatrical projects. Lorimar's motion picture, *Power*, directed by Sidney Lumet, was the first feature film to be edited electronically, using the Montage Picture Processor.

Pacific Video, the largest facility in Los Angeles in 1985, began testing the new electronic systems — the EditDroid, the Montage, and the Ediflex — for use on film-originated television programming. President Emory Cohen created a film environment in a video post facility in an effort to streamline the new electronic process for traditionally film-oriented productions: he coined the term "the Electronic Lab."

1986

✦ Television and some films begin using the new nonlinear systems. ✦ The Montage Computer Company is closed and re-formed. ✦ BHP's TouchVison and CMX's 6000 debuts at the SMPTE show.

Sales slow, and development costs high, heads of both Montage and The Droid Works are troubled. In spite of great excitement over their products, the president and principal owner of Interscope Communications, Ted Fields, announced in March that the Montage Computer Corporation would be sold off to liquidators. Similarly, Droid Works Chairman of the Board Doug Johnson announced the departure of TDW president Robert Doris. The demise of the Montage was ironic in light of the activity on the roughly 29 systems in the field: Alan Alda had switched from editing *Sweet Liberty* on film to the Montage; director Susan Seidelman was planning on using the system for *Making Mr. Right*, and rumors from London said Stanley Kubrick was planning to use the system for *Full Metal Jacket.*

NAB '86 did not see the Montage; however, improved EditDroids and SoundDroids were demonstrated on newer and less expensive SUN/3 computers. By May, the Montage technology had been purchased at auction for $700,000 by a New York businessman, Simon Haberman.

By June, The Droid Works' board had appointed a new president to the company, (the first president of CMX) William Butler.

In August a new company was formed from the remains of the Montage Computer Corporation. Called the Montage Group, Ltd., the new company announced a new smaller Montage II system that would cost $150,000 *less* than their original system, and promised that it would be shown in prototype at NAB '87.

Ediflex boomed in the television rental market. By August, Cinedco announced that all 21 Ediflex systems were in use and that they were turning away business. Lorimar-Telepictures, under vice-president of post production technical services Chuck Silver, pioneered the use of the Ediflex, as well as other videotape-based systems, for use in network television. Universal was also experimenting with electronic post under vice president of post production Jim Watters: Universal's Alan Alda-directed feature *Sweet Liberty,* cut first on film, went through additional cuts on the Montage, following the system's success on Lumet's *Power.*

At the SMPTE show in October, CMX introduced its *CMX 6000* — controlling only laserdiscs and no videotape. It was extremely fast and was the first system with a "virtual master," the ability to roll a "simulated" version of the edited master in all directions and at all speeds, without recording to tape.

CMX 6000 advertisement,
circa Winter 1986

Also at the show was BHP, Inc., the manufacturer of Bruce Rady's *TouchVision* system, now controlling 9 or 12 VHS tape machines instead of 3/4" tape, and featuring a unique touchscreen for control. Notably absent from the show was The Droid

BHP TouchVision advertisment,
circa Winter 1986

Works' *SoundDroid*. Two other products of interest shown were Kodak's prototype of their new film numbering system, *Datakode*; and Cinesound International's *Lokbox* — for interlocking a videotape machine with a film synchronizer. Both would aid in the electronic post production of film.

1987

✦ The Droid Works is closed. ✦ Work begins on the E-PIX, the EMC and the Avid/1.

By the end of January, Lucasfilm announced the closing of The Droid Works, and that no more Droids would be sold. No SoundDroids were ever manufactured beyond the prototype, and only 17 EditDroids were made. The post-production division of Lucasfilm, Sprocket Systems, absorbed the remaining Droids and continued in-house development of the EditDroid with a new development team.

NAB '87 saw the new Montage Group demonstrating their Montage II in prototype.

A software designer on staff at Adcom Electronics, a Toronto-based equipment distributor and manufacturer, wrote a program that would help some filmmakers edit their film project on videotape. Shawn Carnahan, the designer, created a database for establishing a relationship between film edge

numbers and timecode, and Adcom sold the product as "Transform LM." Soon, Carnahan began to add machine control and video switching to the product. He called the system the *E-PIX*.

Desktop computers continued to become more powerful, memory continued to increase in density, and applications became more sophisticated.

Bill Ferster, president of West End Film in Washington DC, had grown increasingly weary in the videotape assemblies of his company demo reels each year. He designed a PC-based editing system that accommodated his needs, storing images digitally on hard disk, and playing them back at 15fps. He decided to sell his seven-year-old animation/graphics company.

At around the same time, William Warner, a marketing manager at Apollo computers outside of Boston, had grown frustrated with the time and cost of the online he endured for the preparation of corporate projects. He, too, began developing an editing system; this one on the Apollo workstation — a high-powered professional system designed for computer graphics applications.

1988

✦ The Montage wins an Oscar. ✦ E-PIX debuts at NAB show; the EMC debuts at SMPTE.

© A.M.P.A.S.®

The Montage Picture Processor won the 1987 Academy Award for Scientific and Technical Achievement — for its use in cutting films such as *Full Metal Jacket, Power, Sweet Liberty,* and *Making Mr. Right.*

By the NAB show in Las Vegas, two new systems were demonstrated in prototype. Adcom debuted its disc-based system, the E-PIX, after eight months of development. Carnahan had designed the system after examining the problems of the major systems in 1987: he did not want a huge bank of peripheral equipment, and ODC disc mastering was still somewhat expensive and difficult to come by. He wanted a stand-alone system that was not tailored to any single market, but generalized to many high-end markets.

William Warner arrived to the NAB show with no intention to market or sell his system, only to get suggestions for improvements. He set up two of his prototypes in a hotel room; the system was called the *Avid/1* and ran on the Apollo computer with a fixed hard disk, played video at 15fps, and had no audio capabilities. Still, interest in the digital system was high, and Warner absorbed the information given him: the system had to run at 30fps, video image had to be better quality, it needed audio tracks, and storage quantity was a big concern.

After NAB, Warner left Apollo and created Avid Technology with venture capital from the Boston investment community. Apple approached Avid, and by the end of the year the prototype systems was ported to a Macintosh computer, video was moved to magneto-optical discs for storage, and playback was at 30fps.

E-Pix advertisement, circa Fall 1989

In July, with $6 million from the sale of his West End Film to Pansophic, Bill Ferster financed a new company to manufacture and market his digital nonlinear editing system running on a personal computer. Ferster called it the Editing Machines Corporation, and the system was called the EMC^2; he had moved the storage media from hard disks to magneto-optical drives. The EMC2 had its debut at the SMPTE show in New York. The system price was under $30,000.

1989

✦ In January, EMC has shipped the first digital systems. ✦ The Avid/1 debuts, and the company begins taking orders.

At the NAB show, Avid Technology began to take orders based on their prototype of the Avid/1, now with full editing capabilities, although no systems had yet been manufactured. At the SMPTE show, Avid demonstrated a second channel of audio; plans were to ship the beta systems by year's end. By December 29th, the first Avids were shipped.

1990

✦ LucasArts shows a new editDROID. ✦ Montage plans to create a fully digital version of their system. ✦ Steady improvements are made in all digital systems.

February, the EMC^2 demonstrated 30fps digital video, and began using the C-Cube compression technology in order to increase resolution.

Early in the year, Lucasfilm unveiled a new company structure and a new *"editDROID."* LucasArts Editing Systems is the manufacturer and developer of the improved system; it was released after a three-year hiatus for re-design; 18 new systems had been built and 12 more were under construction at a plant located on Lucas' Skywalker Ranch. The system was announced as only rentable for its first year.

At the SMPTE show, the Montage Group demonstrated a digital-tape hybrid system, called the *Montage II-H*. This hybrid would lead the way to a fully digital system, expected to be shown in prototype at the next NAB show.

In London, Paul Bamborough, a founding partner of Solid State Logic (SSL), the professional audio console company, begins work on a new system. He had been considering improving editing sytsems since the early 80s, but had felt the EditDroid did about as much as could be done with existing technology. By the late 80s, Bamborough left SSL, and began work on a new graphical system, primarily for film work, that would use compressed digital video. His intention was to make a system that was both simple to use, and more playful than traditional editing systems. He joined with some friends of his who had experience in engineering and system architecture, and formed O.L.E. Ltd., a partnership to manufacture a new system, to be called the "Lightworks."

A programmer at SuperMac designs a new software editing package to help demonstrate their new video digitizing cards for the Macintosh. Randy Ubillos calls his program "ReelTime." By the end of the year, SuperMac has sold the software to Adobe, who renames it "Premiere."

1991

✦ OLE debuts the prototype of the Lightworks. ✦ Apple releases QuickTime. ✦ Chyron (and CMX) are purchased by Pesa.

Early in the year, a small group of people who left Grass Valley Group in 1990 gathered to form a company to make a new kind of video computer. Calling themselves "ImMIX" — a variation on the Greek word for a

Test Drive
Our Flying Machine.

Non-Linear Editing. The Way You Want It To Be.

Avid brochure, circa Summer 1990

gathering of friends — they were acquired by Carleton and incorporated by February.

At NAB, there were eight familiar nonlinear editing systems, and a handful of new ones: the *Montage III* was shown in prototype as expected — a kind of an upgraded Montage II with DVI-based digital video source material; *editDROID* was demonstrated, after having been used on significant portions of Oliver Stone's *The Doors*; the *Ediflex II* was demonstrated with Panasonic write-once discs instead of the familiar VHS decks. *E-PIX* was demonstrated, as was the *TouchVision*. TouchVision, Inc., Rady's new company, demonstrated a prototype of its DVI-based digital system, called *D/Vision*. The TouchVision had just finished its work on the Madonna documentary, *Truth or Dare*. Chyron/CMX showed their familiar *CMX 6000* after having cut a Bernardo Bertolucci film, *The Sheltering Sky*, and a Paul McCartney concert film during the past year; there was a fair amount of activity at both the *EMC2* and *Avid* booths, both showing prototypes of upgraded systems — more editing features and better resolution images. Prototypes were demonstrated for the *Lightworks* system, the Mac-based *Blade Runner* digital editing systems, and D/FX's *Video F/X*.

In May, Apple Computer Company announced a new format for running digital video in many Macintosh applications — called "QuickTime." With a grant from Apple, the American Film Institute opened their new AFI-Apple Computer Center, dedicated to investigating new technologies for film making and professional applications of the new QuickTime format. Adobe releases version 1.0 of *Premiere*.

By year's end, Chyron was purchased by the Spanish video equipment manufacturer Pesa.

1992

✦ Avid introduces their OMF Interchange format. ✦ Hollywood begins to adopt the digital nonlinear systems in place of the original analog ones. Image quality for offline editing reaches an acceptable trade-off between size and resolution.

As expected, NAB was rich with digital nonlinear systems. A release version of the much-anticipated *Lightworks* system was shown, as were a wide array of products from Avid Technology. CMX privately announced a low-cost digital system based on the 6000, to be called the *CMX Cinema*. Montage again showed a prototype of their *Montage III*. All the principal equipment manufacturers demonstrated upgraded systems, with the single exception of the *editDROID* from LucasArts.

The issue of digital image quality finally began to be seen as the red herring it was — diverting spectators from the serious issues of editing system interface, storage, and price. After a number of years where image quality was *the* issue, all systems demonstrated "reasonable" images at comparable compression rates.

Encore Video, in Los Angeles, made the first significant strides in the use of the *Avid* for episodic television. Paramount Pictures began the first significant use of the *E-Pix* for television and, finally, theatrical film work by the end of the year. The Post Group, also in Los Angeles, became the principal *Lightworks* installation, giving the system some of its first serious trials on television programming.

By the end of the year, it was clear that 1992 had been a watershed year for nonlinear editing, and digital nonlinear systems in particular. Avid had placed over 1000 systems worldwide over the previous few years, successfully introducing the concept of nonlinear editing to the video and broadcast communities that had long been unaware of the available products. Now, too, there were viable competitive systems to the *Avid* being evaluated. Television was now taking the digital offline systems more seriously — and theatrical features, the long-time holdout, moved one more step toward the acceptance of these systems as the future of theatrical post.

1993

✦ Montage III and CMX Cinema, digital versions of established systems, are beta-tested. ✦ Avid Technology goes public. ✦ J&R Film Company takes over the editDROID from LucasArts and begins shifting from Moviolas to digital film systems. ✦ "Online" nonlinear systems are debuted. ✦ Unprecedented numbers of feature films move to digital nonlinear systems.

January saw beta shipments of CMX's digital *Cinema*, the only digital departure from the windowed and mouse-driven environments of other systems. Both *Montage III* and *CMX Cinema* were anxiously awaited for their national release at NAB. Both were introduced more slowly than expected.

Avid Technology, riding high as the leading digital nonlinear vendor, made an initial public offering (IPO) in February, and also announced its purchase of DiVA's *VideoShop* — a Macintosh application software product that is a low cost desktop video system — and the principal competitor to Adobe's *Premiere*. Avid also begins shipping first versions of Airplay, NewsCutter and MediaRecorder products. They enter the corporate market with their Media Suite Pro.

NAB was for the first time filled with software companies, offering a whole range of editing products. New arrivals to the scene included Adobe and SuperMac.

Ediflex announced the impending arrival of their *Ediflex Digital,* one more in the growing list of established nonlinear systems moving into the digital domain; although the system was not shown at NAB, it was demonstrated throughout the Spring and initially released that Summer.

J&R Film Company, the owner of Moviola, began the slow phaseout of their Moviola rental business by acquiring digital nonlinear systems for film rental — a dramatic step that was precipitated by other Hollywood establishment film rental houses (like Christy's) doing the same. It was a landmark move in an industry where the Moviola has been the bedrock of traditional editing equipment for almost 100 years.

By April, J&R took over the ownership and support of the editDROID from LucasArts, thus ending the Lucasfilm connection to the product, began more than a decade earlier. Lucas, in turn, announced a plan of future development with Avid Technology.

The buzzwords of the year involved "online nonlinear systems," with the prototype of ImMIX's *VideoCube*, Avid's *4000* and *8000* series, and an early prototype of Lightworks' *Heavyworks*. There were also many new products targeted towards the industrial and corporate "online" desktop: Matrox *Studio*, Data Translations' *Media 100*, and the German FAST Electronics *Video Machine*. Many manufacturers began to unbundle their products into available software and boards, for end-users to use with their personal computers.

The year was a key for theatrical features moving towards electronic film; after well-watched trials for Oliver Stone (on the Lightworks) and Martha Coolidge (on the Avid), those two systems made the move into feature films and began steady work as accepted tools for Hollywood.

1994

✦ Avid Technology increases massive market penetration and differentiation of products. ✦ CMX Cinema project is ended. ✦ Lightworks' Heavyworks is the first viable digital nonlinear system used for traditional multicamera television programs. ✦ Cinedco, owner of the Ediflex, closes its doors; sells patent rights to Interfilm, Inc. ✦ Many manufacturers begin steering their video products into digital and nonlinear domains while existing nonlinear manufacturers push the image resolution limits with better and better images — further blurring the online/offline distinctions. ✦ EMC is purchased by Dynatech. ✦ Softimage is purchased by Microsoft.

After 5 years in business, Avid reported gross sales of $203 million dollars and profits of 4 million. Their R&D budget exceeds the gross sales of every other nonlinear manufacturer to date. At NAB they make the first public demonstrations of their OMF (Open Media Framework™) Interchange. By December, Avid had purchased Digidesign, a leading digital audio company.

Montage settled into slow adoption of their digital *Montage III*, but continued to aggressively pursue enforcement of their patents concerning digital images in editing systems.

Adcom phased out their E-PIX editing system, and replaced it with a high end digital "online" nonlinear product called *Night Suite*, boasting D1-quality video output.

Sony demonstrated a prototype of their first nonlinear product, called the *Destiny* Editing System. One of the key designers of the Destiny system was Robert Duffy, the engineering designer of the CMX 6000.

1995

✦ Tektronix purchases Lightworks. ✦ Grass Valley signs agreement with Data Translations to distribute the Media 100. ✦ Avid purchases Parallax Software Group, developer of high-end paint systems; and Elastic Reality, a leading developer of image special effects software. A prototype of an AvidDroid is shown at NAB. Avid licenses Ediflex's Script Mimic technology. ✦ Inventors of the Avid and Lightworks products win 1994 Academy Awards for Technical Achievement. ✦ All major networks have broadcast programs directly from nonlinear systems. ✦ Sony stops development of its nonlinear *Destiny* product. ✦ According to an NAB survey, 38% of broadcasters use nonlinear equipment.

In March, Avid acquired Parallax Software, developer of high-end paint and compositing systems *Matador* (2D paint and animation) and *Advance* (compositing and image processing). At the same time, Avid also acquired Elastic Reality, a software developer known for their advanced morphing and image restoration products. At NAB, Avid showed versions of products that run on Macs, PCs, and introduced new Silicon Graphics configurations. They also debuted an SGI Onyx-based online system that incorporates their products. According to company literature, Avid employs 1200 people. Their presence at the NAB show was pervasive.

Announced at the NAB show was the merger between video giant Tektronix (revenues of $1.3 billion in 1994) and Lightworks. As of NAB, Lightworks employed 100 people and had *approx.* 1000 high-end film cutting systems in the worldwide market.

Grass Valley Group announced a Mac-based personal production suite called VideoDesktop. Although the product literature makes no mention of it, the system is the result of an alliance with Data Translation where GVG is able to assemble and sell turnkey Media 100 systems. Grass Valley also shipped their V.1 of their "nonlinear" Sabre System.

A new company arrived at NAB made up of 11 senior staff members from NewTek (makers of the Video Toaster) who call themselves Play, Inc. The hip team demonstrated a prototype of a PC-based, D1 video bus (rather than PCI) post-production system called *Trinity*.

Following NAB, Avid announced a patent infringement suit against Data Translation, designer of the Media 100 and a rising competitor to some of Avid's products. The suit concerned proprietary audio and video manipulation methods internal to the Macintosh.

Ironically, in a year with more than 120 hard disk-based nonlinear-type products, there were more video tape formats introduced than at any previous NAB. The buzz at the show was on MPEG compression, networking systems, and in dedicated broadcast news nonlinear systems.

The year showed a marked increase in development on the PC as a video platform: with the introduction of Microsoft's Windows 95, plus the creation of the Open Digital Media "Language" (OpenDML) consortium, on top of the continued development of PCI-bus plug-and-play video boards, windows-based products had a surge in popularity.

The NAB show itself was in many ways a metaphor for the state of the industry. The giant convention center in Las Vegas had an annex in 1991 (and for the previous few years) dedicated to HiDefiniton Video; by 1992 that was superseded with something called Multimedia World. By 1995 it was as if Multimedia World was beginning to reach adolescence; the convention was split and it was not unlike the "adults" in the main floor, and the "kids" in multimedia world. It perfectly represented the transition from the broadcast realm to the computer realm; from the establishment to the revolution. It was not uncommon to see midlife broadcasters wandering lost and confused among the multimedia booths; and to see youth drunk with their newfound power, but not quite sure what to do with it. More than anything, NAB exemplified the still-enormous knowledge gap between the two worlds, and the few participants who could speak both languages. The corporate giants of the establishment having failed to truly invent the future were forced to grab it by brute force, buying and merging with the new technologies and placing them, like badges, on their corporate jackets.

T H E   M O N T A G E   P A T E N T

Very often, a company does more than build a really great editing system. Sometimes they actually invent something totally new. And when they do this, they usually patent it. Once patented an invention can work to give a company a competitive edge in their product line, or the patent can be licensed to other companies which can give the inventor extra revenue in the form of royalties. Some manufacturers have not patented their original inventions, and so these features have found their way for "free" into the array of products you see at shows: like timelines and rippling edits. A core bit of research was performed at XEROX PARC in the late 1970s — their unpatented inventions are widespread in computer graphics and editing systems worldwide: the mouse, the bit-mapped display, the icon, and so on.

There are many patents that are central to the development of nonlinear editing systems, many very technical and some more theoretical. Among the important patents and inventions worth noting in the history and future of systems are ones like the first nonlinear system (1970, patent number 3,721,757, by Adrian Ettlinger) to ones originally from Cinedco's Ediflex — the script mimic — licensed by Avid in 1995.

But of all the patents with which manufacturers must deal, the most important has been the one for digital picture labels used in editing systems. Patented in 1982 and first demonstrated in the form of the original Montage Picture Processor, Ron Barker and Chet Schuler's work has been hotly contested but never rebuked. And as the Montage Group has evolved over the past decades, this patent has proven a valuable asset to the company, one of the cornerstones of nonlinear editing. Excerpts of the patent's 67 claims are reprinted here for serious students of the industry.

PATENT NUMBER 4,538,188
AUGUST 27, 1985

Inventors: Ronald Barker & Chet Schuler
Assignee: Montage Computer Corporation
Abstract Excerpt: "A video composition apparatus and method select (sic) segments from image source material stored on at least one storage media and denote serially connected sequences of the segments to thereby form a composition sequence. The apparatus and method employ pictorial labels associated with each segment for ease of manipulating the segments to form the composition sequence."

Noteworthy Claims: (quoted from patent documents)

#1   Composing apparatus for selecting segments from image source material stored in at least one storage medium and for denoting serially connected sequences of said segments...

#36   A composing method for selecting segments from image source material stored in at least one storage medium and for denoting serially connected sequences of said segments, and said method comprising the steps of:

- denoting a start and end of each of a plurality of segments of said source material
- identifying each said segment by a pictorial image segment label
- selectively displaying segments of said image source material on a pictorial display means
- assembling at least a plurality of said segment labels into serially connected label sequence
- locating any segment by displaying said pictorial labels in said label sequence and
- displaying the segment identified by a selected label

#41   [In addition], the labeling of each defined segment by two pictorial images, a first label image corresponding to a beginning frame of said segment and a second label image corresponding to an ending frame[.]

Comments: Filed in December 1982, this is the fundamental patent that covers many editing system traits, but most importantly the digital picture labels used in editing ("video composition"). While the patent was designed for the Montage Picture Processor, first released in 1984, the patent goes on to cover clearly (1) picture labels used to identify a source shot or segment and (2) the use of picture labels in a sequential order (like in a timeline) that identify cut segments; such that when you move the picture labels around you change the order of the shots in the edited sequence; and (3) the head and tail label method for identifying a shot.

While the patent was rarely mentioned in the early years of nonlinear editing, it arose as central with the advent of the digital systems in the late 1980s. While a number of graphics and nonlinear manufacturers have licensed the use of picture labels from Montage over the past years, there has been some contention that the feature is not valid due to "prior art" — that is, that some products prior to 1982 demonstrated the use of digital picture labels in composing (*e.g.* the Asaka patent, #4,612,569). Regardless, the patent has never been successfully challenged and most manufacturers of digital systems have licensed the features.

T H E A T R I C A L   F E A T U R E S
AND THE FIRST DECADE OF NONLINEAR EDITING

For many people, the ultimate adoption of electronic editing for film was determined by its use for theatrical feature films. Slowly over the course of the years from 1985 through 1995, more and more projects have been edited partially or entirely on this new equipment.

The following list follows the introduction of electronic systems for theatrical film work. In many ways this kind of listing is misleading, as many systems have been cutting episodic television series and movies-for-television, both of which often cut negative from the editing system. In fact, television movies (like 1988's *LONESOME DOVE*, cut on the CMX 6000) have more in common with theatrical releases like *THE GODFATHER III* (cut on the Montage Picture Processor) than they do with most other television programming. **This list is by no means the entirety of films edited electronically.** Particularly in the years beginning with 1993, the number of films cut electronically has soared.

Since some systems have cut hundreds of low-budget, foreign, or video-released films, the main listing (in **bold** typeface) will show only films with budgets over three million dollars. For some scale, the *first* film any system cut is also included (in *italics*), regardless of the budget.

Year	Film Title	Director	Editor*	System
1985	**POWER**	Sidney Lumet	Andrew Mondshein	*Montage*
1986	**SWEET LIBERTY**	Alan Alda	Michael Economou	**Montage**
	MAKING MR. RIGHT	Susan Seidelman	Andrew Mondshein	**Montage**
	The Patriot	*Frank Harris*	*Richard Westover*	*EditDroid*
1987	**FULL METAL JACKET**	Stanley Kubrick	Martin Hunter	**Montage**
	Perfect Victims	*Shuke Levy*	*Jon Braun*	*CMX 6000*
	Garbage Pail Kids	*Rod Amateau*	*Leon Carrere*	*Ediflex*
	Mine Field	*Jean-Claude Lord*	*Yvles Lang Louis*	*TouchVision*

(1987 Academy Award, Scientific & Engineering achievement to MONTAGE)

Year	Film Title	Director	Editor*	System
1988	**TORCH SONG TRILOGY**	Harvey Fierstein	Nicholas Smith	**Montage**
	BIG TIME	Tom Waits	Glen Scantlebury	**Montage**
	COOKIE	Susan Seidelman	Andrew Mondshein	**Montage**
1989	**FAMILY BUSINESS**	Sidney Lumet	Andrew Mondshein	**Montage**
	EDDIE and the CRUISERS 2	Jean-Claude Lord	Jean-Guy Montpetit	**TouchVision**

* The first editor or the editor responsible for the electronic editing.

Year	Title	Director	Editor	System
1990	THE SHELTERING SKY	Bernardo Bertolucci	Gabriella Cristiani	CMX 6000
	GRAFFITI BRIDGE	Prince	Rebecca Ross	CMX 6000
	GET BACK	Aubrey Powell	Michael Rubin	CMX 6000
	ROCKY V	John Avildsen	Michael Knue	TouchVision
	THE OBJECT of BEAUTY	Michael Lindsay-Hogg	Ruth Foster	Ediflex
	THE DOORS	Oliver Stone	David Brenner	EditDroid
	ONCE AROUND	Lasse Hallstrom	Andrew Mondshein	Montage
	GODFATHER III	Francis Coppola	Barry Malkin	Montage
1991	TRUTH OR DARE	Alex Keshishian	Barry A. Brown	TouchVision
	HARLEY DAVIDSON and the MARLBORO MAN	Simon Wincer	Corky Ehlers	CMX 6000
	DECEIVED	Damian Harris	Neil Travis	Montage
	LEAVING NORMAL	Ed Zwick	Victor duBois	Ediflex
	ALL I WANT for CHRISTMAS	Rob Lieberman	Dean Goodhill	E-PIX
	Critters 3	*Kristine Peterson*	*Terry Stokes*	*E-PIX*
	Let's Kill All the Lawyers	*Ron Senkowski*	*Christa Kindt*	*Avid*
	High Strung	*Roger Nygard*	*Tom Siiter*	*EMC2*
1992	KAFKA	Steven Soderbergh	Steven Soderbergh	EditDroid
	MEDICINE MAN	John McTiernan	Michael Miller	EditDroid
	WIND	Carol Ballard	Michael Chandler	EditDroid
	PATRIOT GAMES	Phillip Noyce	Neil Travis	Montage
	B. STOKER'S DRACULA	Francis Coppola	Nicholas Smith	Montage
	DISTINGUISHED GENTLEMAN	Jonathan Lynn	Tony Lombardo	CMX 6000
	STREET KNIGHT	Albert Magnoli	Wayne Wahrman	CMX 6000
	ARMY OF DARKNESS	Sam Raimi	Bob Murawski	EMC2
	KALIFORNIA	Dominic Sena	Martin Hunter	*Lightworks*
1993	LOST IN YONKERS	Martha Coolidge	Steve Cohen	Avid
	NEEDFUL THINGS	Fraser Heston	Rob Kobrin	Avid
	THE FUGITIVE	Andrew Davis	Dean Goodhill, *et al.*	Avid
	THE GETAWAY	Roger Donaldson	Conrad Buff	Avid
	ANGIE	Martha Coolidge	Steve Cohen	Avid
	LOVE AFFAIR	Glenn Gordon Carron	Bob Jones	Avid
	A BRONX TALE	Robert DeNiro	David Ray	Avid
	TRUE LIES	James Cameron	Conrad Buff, *et al.*	Avid
	TWO BITS	James Foley	Howard Smith	Avid
	NAKED GUN 33 1/3	Peter Segal	Jim Symons	Avid
	WOLF	Mike Nichols	Sam O'Steen	Avid
	YOUNGER and YOUNGER	Percy Adlon	Suzanne Fenn	Avid
	THE THING CALLED LOVE	Peter Bogdonovitch	Terry Stokes	E-PIX
	THE CONEHEADS	Steve Barron	Paul Trejo	E-PIX
	SLIVER	Phillip Noyce	Richard Francis Bruce	Montage II
	HEAVEN AND EARTH	Oliver Stone	David Brenner	Lightworks
	IRON WILL	Charles Haid	Andrew Doerfer	Lightworks
	BEVERLY HILLBILLIES	Penelope Speeris	Ross Albert	Lightworks
	MRS. DOUBTFIRE	Chris Columbus	Raja Gosnell	Lightworks
	INTERSECTION	Mark Rydell	Mark Warner	Lightworks
	WHITE FANG 2	Ken Olin	Elba Short	Lightworks
	NATURAL BORN KILLERS	Oliver Stone	Hank Corwin	Lightworks
	PELICAN BRIEF	Alan Pakula	Tom Rolfe	Lightworks
	CLEAR / PRESENT DANGER	Phillip Noyce	Neil Travis	Lightworks

* The first editor or the editor responsible for the electronic editing.

EIGHT SECONDS	John Avildsen	Doug Seelig	Lightworks
BLUE CHIPS	William Friedkin	Robert Lambert	Lightworks
DISCLOSURE	Barry Levinson	Jay Rabinowitz	Lightworks
BLOWN AWAY	Stephen Hopkins	Tim Wellburn	Lightworks
SPEED	Jan De Bont	John Wright	Lightworks

1994	ANGELS IN OUTFIELD	William Dear	Bruce Green	Avid
	FORGET PARIS	Billy Crystal	Kent Beyda	Avid
	DOLORES CLAIRBORNE	Taylor Hackford	Mark Warner	Avid
	RADIOLAND MURDERS	Mel Smith	Paul Trejo	Avid
	READY TO WEAR	Robert Altman	Geraldine Peroni	Avid
	TOMMY BOY	Peter Segal	William Kerr	Avid
	DROP ZONE	John Badham	Frank Morriss	Avid
	BRAVEHEART	Mel Gibson	Steve Rosenblum	Lightworks
	PULP FICTION	Quentin Tarantino	Sally Menke, et al.	Lightworks
	DIE HARD: With Vengeance	John McTiernan	John Wright	Lightworks
	NELL	Michael Apted	Jim Clarke	Lightworks
	OUTBREAK	Wolfgang Peterson	Neil Travis, et al.	Lightworks
	CONGO	Frank Marshal	Anne V. Coates	Lightworks
	FRENCH KISS	Lawrence Kasdan	Joe Hutshing	Lightworks

(1994 Academy Award, Scientific & Engineering achievement to AVID)
(1994 Academy Award, Scientific & Engineering achievement to LIGHTWORKS)

1995*	BRIDGES OF MADISON COUNTY	Clint Eastwood	Joel Cox	Avid
	WHILE YOU WERE SLEEPING	John Turteltaub	Bruce Green	Avid
	SPECIES	Roger Donaldson	Conrad Buff	Avid
	VIRTUOSITY	Bret Leonard	Rob Kobrin, et al.	Avid
	THE NET	Irwin Winkler	Richard Halsey	Avid
	SHOWGIRLS	Paul Verhoeven	Mark Goldblatt	Avid
	NIXON	Oliver Stone	Brian Berdan, et al.	Avid
	JUMANJI	Joe Johnston	Robert Dalva	Avid
	ASSASSINS	Richard Donner	Richard Marks	Avid
	SABRINA	Sydney Pollack	Fredric Steinkamp	Avid
	BATMAN FOREVER	Joel Schumacher	Dennis Virkler	Lightworks
	NINE MONTHS	Chris Columbus	Raja Gosnell	Lightworks
	WATERWORLD	Kevin Reynolds	Peter Boyle	Lightworks
	MISSION: IMPOSSIBLE	Brian De Palma	Paul Hirsch, et al.	Lightworks
	MATILDA	Danny DeVito	Lynzee Klingman	Lightworks
	BOGUS	Norman Jewison	Stephen Rivkin	Lightworks
	VAMPIRE IN BROOKLYN	Wes Craven	Patrick Lussler	Lightworks
	NO FEAR	James Foley	David Brenner	Lightworks
	HOME FOR THE HOLIDAYS	Jodie Foster	Lynzee Klingman	Lightworks
	LAST DANCE	Bruce Beresford	John Bloom	Lightworks
	CASINO	Martin Scorsese	Thelma Schoonmaker	Lightworks

Although equipment manufacturers won't generally advertise it, many of these theatrical features were cut partially on electronic systems and partially on traditional workprint. The electronic cutting room, like the original film room, might have any combination of equipment: flatbeds, uprights, synchronizers, log books . . . similarly, there may be one or more electronic systems as well. There is no shame in only partially contributing

* by mid-1995

to the process that post-produces a feature film — any way to efficiently, quickly, and creatively pour through 500,000 feet of film is an advancement to the film industry. A handful of directors have been using videotape to assist in production and post production for some years now, and even prior to 1985. Inexpensive linear tape systems have been employed to rough-together scenes before cutting film. On rare occasions, entire movies have been edited linearly, on tape, and then conformed manually back to film.

Although all of the projects listed above are technically "theatrical" feature films, many of the lower budgeted films (like the countless ones not shown here) had only a contractual film release and then went directly to videotape distribution (called *direct-to-video* films). Regardless, all cut negative either directly from the systems or from a conformed workprint made (at least in part) from lists output by an electronic editing system.

This listing concludes with the films cut in 1995 as the principal equipments used in feature film work, the Avid Film Composer and the Lightworks, have both been distinguished with Academy Awards, aproximately 10 years after the first films were cut electronically on the Montage. Clearly, the caliber and the quantity of films has grown substantially, and electronic film editing has today crossed over into the mainstream of Hollywood.

[For the most complete records of projects editing on any brand of nonlinear equipment, it is suggested that you contact the manufacturer directly.]

INDEPENDENTS AND NONLINEAR

"Why don't more independent film makers and experimental films use nonlinear editing?"

In one sense, independents have as much to gain (if not more) than more commercial, higher budgeted productions. At a creative level, the flexibility and speed are paramount and could change a year-long project into one of perhaps a few months. Documentaries and experimental films usually have more creative alternatives to exhaust than do typical Hollywood scripted projects.

One reason these projects have not jumped on nonlinear editing is the nature of the early nonlinear systems. With "loads" of only a few hours at most, many systems had some difficulty with the organization and the efficient editing of non-scripted material. The highly variable way in which the material needed to be joined and rearranged made cutting problematic on many earlier systems. Even the digital systems often have difficult "load" requirements (in particular, those systems utilizing MO disks for source) that need to be overcome. Still, this barrier has been steadily reduced through advances made to the systems and better economics of large gigabyte storage options.

Secondly, and probably more significant, are reasons of money. Nonlinear editing has been adopted easily by Hollywood television and films because of the potential for cost-savings. With the cost of money, each month saved in post production translates directly to dollar savings. With major productions, there are immediate benefits to finishing sooner — release dates, contractual delivery obligations — as well as the premiums paid to directors and editors each week. If an editor makes two to five thousand dollars A WEEK, simple time savings also translate into huge labor savings.

Clearly, for larger budget projects with an expensive team, time equals a lot of money. The price paid to edit electronically can be dramatically recouped by the cost savings associated with time. But for independent projects, this is not the case. There is often no delivery date or any kind of contractual finish time that must be met. The cost of directors and editors is usually very low. Weeks saved in post-production amount to little direct cost-savings. At the relatively high price of electronic nonlinear, it is difficult to justify the costs for low-budget independents. Electronic nonlinear systems originally rented for about $2,500 per week. Due in part to competition, prices on these systems have dropped about 30% in the past couple of years. But keep in mind that the equivalent film equipment can be rented for a few hundred dollars per month.

Still, saving time is saving time. If time has any premium associated with it — if a production simply wants to work quickly with the added benefits of high flexibility and interactivity — nonlinear alternatives are still attractive. If directors or editors are perhaps donating or investing their time, the incentive to finish soon still exists.

Finally, there is the issue of digital storage. Very often in low budget productions, the film project may temporarily run out of money. When this happens, filming stops, teams are dispersed for an indeterminate amount of time, and for the most part, editing would cease. With film or video, this is no big deal: you release the edit bay, you pack up tapes or film into boxes, and wait until more money comes in. But for digital nonlinear productions, the digitized source material cannot be stored on the system's hard disks during this break without significant costs. If the source material is simply deleted from the system, while it can be re-digitized when the post-production resumes, there will be a huge penalty in terms of time and effort. On both ends, it just isn't worth the trouble and cost if there is any possibility that the production will be halted in midstream.

Today, with the digital nonlinear systems finally matured, many nonlinear systems are affordable — to buy and to rent. Until recently, all electronic nonlinear systems were priced in the hundreds of thousands of dollars. Even as prices dropped and digital systems began to evolve, rental costs were still beyond the reach of many kinds of productions. But the 4th generation is signaled by a dramatic decrease in costs. There are now systems at every price point (from the desktop size up to dedicated feature film-oriented systems) and any project should be able to locate rental costs for systems comparable to or less than most film equipment.

There is no replacement for the simplicity and the physicality of working on film. If the lifestyle and the tactile nature of film are the appeal, as is the case with many filmmakers, the process and labor IS the joy of filmmaking, and editing electronically may offer little benefit.

Today, there are more and more places where independents can turn to find this equipment: universities own nonlinear systems; smaller facilities are purchasing equipment and sometimes offer off-peak sessions at excellent rates. As the glut of available systems increases, it is the customer who benefits most. Prices will continue to drop and information about the systems will continue to increase. It is simply impossible today to attend any independent film festival (Sundance, Telluride, *etc.*) and not see copious nonlinear systems mentioned in the screen credits. After less than a decade of watching the advent of nonlinear systems from the wings, independents are now fully empowered to utilize these technologies.

Alternate Methods for Independent Film Making

Film post production can be expensive. You have the costs of film stock, lab, prints — and they are all upfront costs — which makes it kind of scary for cost-conscious individuals. Shooting on videotape is considerably less expensive (tape stock is relatively cheap, there is no lab, *etc.*)

Editing on film is simple, nonlinear, and slow. Editing in video, linear or nonlinear, is equipment-intensive, but can be fast. In the past few years, some short projects and documentary-type projects have shot on videotape and edited on videotape. But linear tape editing is mind-numbingly slow for these projects.

Enter new technology and nonlinear editing. Today is it not only possible but increasingly attractive to shoot on videotape, edit nonlinearly, and release on film. Yes, FILM. Special facilities, like Image Transform in Los Angeles, do impressive transfers of video to 16mm or 35mm film. The quality of the film print is directly related to the quality of the master videotape. HiDef video produces wonderful 35mm film prints. However, even modern *consumer* video technology — like S-VHS and Hi-8 — produce almost-professional quality, that if post-produced correctly, can yield acceptable 16mm prints, suitable for theatrical presentation at festivals, for example. Newer digital camcorders (*e.g.* the DVC camcorder) are coming down in price so fast that many will be within the reach of consumers/prosumers in the coming years. Many Academy Award nominated documentaries and shorts were shot and edited on videotape and transferred to film for Academy consideration. The 1991 winner for best documentary short — *Deadly Deception: General Electric, Nuclear Weapons, and our Environment* — was shot on many different video formats. 1995's highly-awarded film *Hoop Dreams* was shot on both Beta and Beta SP videotape over a five year period by director Steve James — then transferred to 16mm for the Sundance Film Festival, and finally to 35mm for Fine Line Film's theatrical release.

TECHNIQUES

Image Transform, like other facilities that transfer tape to film, recommends that for the best transfer, videotape be component video. Although neither S-VHS nor Hi-8 is component, they are both *pseudo-component* (or Y-C), and can produce exceptional images. If you maintain the component signal throughout post, you will have a suitable master tape for the transfer. One possible flow chart for this is on the following page:

Shoot picture and record audio on Hi-8 or S-VHS pro-sumer camcorder. If budget allows, also record audio on peripheral tape, *e.g.* 1/4" or DAT.

Dub source videotapes to timecoded Beta SP cassettes. These tapes are now the official source tapes.

Digitize Beta SP tapes into a nonlinear system. Edit.

After offline, either output the EDL for an online of the Beta SP source tapes, or use the nonlinear system as a Beta SP controller and auto-assemble the Beta SPs tape-to-tape. If the digital images are of high enough quality, record the picture and audio directly to Beta SP.

For theatrical presentation, "transform" Beta SP master tapes to 16mm.

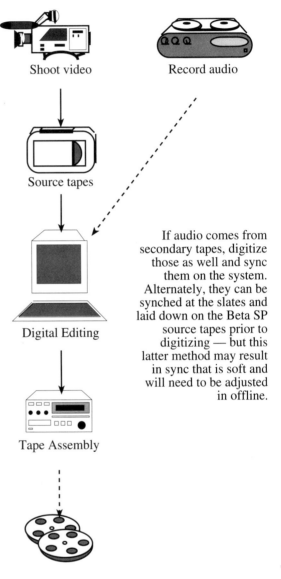

Shoot video

Record audio

Source tapes

Digital Editing

Tape Assembly

If audio comes from secondary tapes, digitize those as well and sync them on the system. Alternately, they can be synched at the slates and laid down on the Beta SP source tapes prior to digitizing — but this latter method may result in sync that is soft and will need to be adjusted in offline.

SPECIAL NOTES

AUDIO: There are two ways to record audio for these kinds of projects. First, you could use the audio as recorded directly onto the videotape via either the on-camera microphone or a remote microphone attached through a jack on the camera. This is *single-system* audio. The audio quality will be limited by the quality of the videotape, the camera, and the microphone — few of which you have much control over. On the other hand, this audio will always be in sync with the picture.

Second, you could record the audio separately with either 1/4" or DAT recorders, as is done on film projects. This is *double-system* sound and might be more difficult to post produce. Its quality will be very good, but because the cameras are consumer models, there is no crystal sync to lock the speed of the audio and video tapes. On a linear editing system, this issue would be virtually insurmountable. But with nonlinear editing, the ability to control and ripple audio makes this viable.

Digital nonlinear editing systems generally allow for CD-quality (44.1 KHz sampling, 16-bit) audio to be recorded into the system during digitizing. If this is done, no tape-to-tape assembly is necessary for sound. The audio tracks can be laid down to a master videotape (Beta SP, 1", D2, for example) in real time. This will dramatically save time in online or in any auto-assembly mode. Also, since Beta SP tapes have high quality analog audio tracks (often mistaken for digital), this will provide for excellent sound prior to any further audio post that might be determined necessary. For very small projects, offline production tracks are sufficient.

TIMECODE: Until there is some kind of timecode that can run on consumer videotapes and be utilized by editing systems, there will always be an additional step in post production to dub the original tapes to high-quality tape stock with timecode, for example, Beta SP.

When either the Hi-8 or S-VHS tapes have timecode — or — when the digital video image in the "offline" nonlinear editing system is of sufficient picture quality, no dubs will need to be made. With timecode, following the offline edit, a tape-to-tape assembly to Beta SP will provide the edited master; with high-quality digital video, you can record to Beta SP directly from the editing system, much as can now be done with digital audio.

SLATES: No slates are necessary *unless* you are recording audio separately and want to sync it in post (whereby you might want both head *and* tail slates to help quantify any drift between the two machines). If time and budget allow, however, slates are a good idea for recording notes and shot information at the head of each take.

S-VHS and Hi-8: Although the quality of S-VHS video is not quite suitable for network broadcast, the quality of S-VHS's luminance signal is. In other words, if you shoot in S-VHS and transfer just the luminance (or black and white) portion of the picture to the edited master, you can have virtually broadcast quality (albeit monochrome) video. Certain types of productions can sometimes take advantage of this.

Hi-8 cameras offer virtually the same degree of quality, but with a camera that is a fraction of the size and weight. Hi-8 cameras are ideal for shooting inconspicuously, where S-VHS are not. On the other hand, S-VHS cameras, because of their size and shoulder positioning, can be steadier.

Actually, consumer video of all types often end up being broadcast on television: videos of newsworthy events, certain television programming like the popular *America's Funniest Home Videos*, and so on. If material is worthy, video quality will generally not keep it from being broadcast.

PRODUCTION: There are certain types of shooting situations that can accentuate the inadequacies of home video. Fast moving camera pans and zooms will "judder" in the transfer to film — a kind of stuttering or jumping of the image that can be distracting. Also, certain areas of high detail can be lost. Although professional lighting or outdoor lighting produce better results, these type of production luxuries are not always available. The more film-like the production set-up and lighting, the better looking this transfer to film. TIPS: steady camera moves (as with a Steadicam Jr.) and tripod use (especially with a fluid-head tripod) help dramatically. Also, shoot with manual focusing and aperture settings.

As nonlinear (and desktop) video proliferates, universities, high schools, and many individuals will have the means to independently produce projects with unprecedented quality at minimal cost. In some way, the entire history of filmmaking and broadcast communications has been leading up to the events these technologies deliver.

CHAPTER 3

FUNDAMENTALS

OFFLINE AND ONLINE

Electronic nonlinear editing has often been referred to as "offline" editing. And newer systems are said to perform "online" editing. But what does this mean? What is the difference between offline and online?

OFFLINE EDITING: The preliminary editing on copies of source material. It is a cost-effective way for editors (along with directors and producers) to make creative decisions and revisions.

In offline editing, the source elements used are always duplicates of the original sources because of the risks and costs involved with handling and shuttling the original tapes. Also, the costs of broadcast-quality equipment are very high per hour; using that equipment to make slow and creative choices would be prohibitively expensive.

The goal of offline is to select and assemble the best performances while seamlessly joining pictures and sounds together. The creative picture editor is primarily concerned with the craft of selecting and matching pictures and dialogue, and then making revisions. The only video effects that the editor is concerned with are simple ones — cuts, wipes, and dissolves. Anything more than this is rare. The product of the offline session is a standardized EDL.

Except for video frame accuracy, everything is allowed to be compromised in order to minimize expense and maximize efficiency. Compromises might include image quality, optical effects, and the lack of final elements such as music, audio effects, and titles.

Offline edit bays typically include three to five 3/4" videotape machines, with one being designated as the record deck, and the rest as sources. There will also be a keyboard-style edit controller, and somewhat inexpensive audio mixers and small video switchers. Aside from B-rolls (duplicates of whole or part of existing source reels), there is usually a single 30-minute copy of each 3/4" source tape—never longer than 60 minutes — often with an inscribed or "burned-in" timecode window in the picture area.

Although the window usually obscures part of the source image, it gives the editor an easy numeric reference to each frame, whether on source tapes or after recording to a master tape. The advantages are considered to outweigh the disadvantages.

04:12:22:08

ONLINE EDITING: The efficient re-creation of events from the offline EDL, to produce the finished product.

The elements used are the videotape masters, traditionally 1" tape, but also Betacam, D1 or D2 digital videotapes. The goal of online is to complete and augment the offline decisions with optical effects, color correction, titles, audio adjustments; specifically, to complete the finished video product to broadcast specifications.

The online editor may not make creative changes to the offline cut. The craft here is to build the program based on the EDL "blueprint" and supplement it with the available broadcast-quality tools. In addition to the assembly from the offline EDL, there are manual and automatic manipulations of the list (cleaning, formatting for broadcast, etc.) that the online editor may perform.

Although the assembly portion of online is not particularly creative (it is somewhat automated by the edit systems), online creativity is largely a function of creating or "building" special video effects for inclusion in the production. Often, sequences that are "effects intensive," meaning they are being composed of mostly-manipulated images, can only be edited in online — no offline is practical.

Online equipment is expensive. Because of this, it is often shared by many bays. The online bay usually includes an edit controller similar to those found in an offline bay, but connected to large video switchers, image manipulation (graphics) devices, and character generators (for titling). With larger budgets, audio is finished separately (called *sweetening*) in facilities designed for that purpose.

IS FILM EDITING "OFFLINE" OR "ONLINE"?

Offline and *online* are strictly videotape editing terms. However, any editing session that produces an EDL for later assembly of final elements is in some sense "offline." That is why the electronic nonlinear systems are all usable as offline systems.

In the film world, the closest parallel to online/offline is editing workprint and cutting negative. The film editor makes creative decisions and builds an edited reel from positive duplicates of the original negative. The reason for this has little to do with cost, but rather the fact that you don't want to handle your original negative — film editing being a sort of destructive-then-constructive process. Besides, you need a positive image to view material.

Once "offline-types" of creative decisions are finalized on workprint, that film is given to a negative cutter, who performs the rough equivalent of "online." The negative cutter does not have a paper list (like an EDL), but rather the cut duplicate print from the negative. Using that duplicate print as a guide, the negative cutter reads the numbers off the edges of the film to locate and assemble the original negative. It is a slow and dangerous process. Any mistake in the negative cutting can be disastrous. Negative cutters, like online editors, are not concerned with creative decisions about the edit, only in the faithful re-creation of the editor's decisions.

Since electronic nonlinear systems often do not involve workprint, but still end up desiring a negative cut, the only way to inform the negative cutter of what to assemble is a printed list — a film-version of an EDL — called a "negative assembly list." This list has entries for each shot's in-points and out-points, in a format very similar to the video-style EDL. Although they are not intended to replace a cut list, videotape dubs of a final cut are usually given to negative cutters to provide a visual reference. Special electronic-controlling synchronizers, for example the LokBox, allow for videotape cassettes with VITC to track along with the film.

New Definitions

Today, these terms get mixed-up and confused. Except when qualified by the word "traditional," *offline* and *online* have new specific meanings.

The words have to do precisely with the **completion-degree** of the **output product** and only incidentally with the output image quality:

ONLINE is the *production of the completed product* — with all associated integration and finality. Thus the image and audio quality inherant is that which is required for delivery of the project. Online image qualities vary depending on the needs of the product's recipient. For television broadcast, online product has "broadcast quality" signals and "high" quality pictures. But if a lesser quality is required for delivery, as for personal use or many non-broadcast uses, "online quality" will be totally different.

OFFLINE is the *production of any interim stage of the product*, and tends to have an image or audio quality that is less than that for online, *but not by definition*. It could produce product with the online image quality, but may not be fully-integrated with all other elements (titles, effects, *etc.*). It might have only a tiny bit less quality; it might have considerably less quality. Since quality has many factors (frame rate, image/pixel resolution, color resolution, aspect ratio, *etc.*) any of these might be compromised to decrease the cost of the creative edit session, where all that is required is *decision-quality video*; that is, resolutions good enough to make creative decisions.

T I M E C O D E

In the early 1970s, videotape editing changed from control track editing to using timecode. An 8-digit timecode defined and electronically identified each video frame with a unique "code" number broken down into. . .

HOURS: MINUTES:SECONDS:FRAMES

01:00:00:00

Timecode made videotape editing more efficient than it had been before. In addition to speeding up the editing process, it was considerably more "frame accurate" than the earlier and cruder control track editing. It also allowed any edit to be previewed and repeated.

Later, computer-controlled editing systems were introduced that could read timecodes, identify edit points, and store lists of these edit decisions.

"Drop Frame" vs "Non-Drop Frame" Timecode

For our current discussion, videotape plays at 30 frames per second (30fps). Videotape timecodes, therefore, count from frame :00 to frame :29 before rolling over to the next second —

01:00:00:00
01:00:00:01
01:00:00:02
01:00:00:03
01:00:00:04
01:00:00:05
01:00:00:06
01:00:00:07
01:00:00:08
01:00:00:09
01:00:00:10
01:00:00:11
01:00:00:12
01:00:00:13
01:00:00:14
01:00:00:15
01:00:00:16
01:00:00:17

Unfortunately, for a variety of electronic and technical reasons, videotape only THEORETICALLY runs at 30fps; in reality it runs at 29.97 frames per second. So, even though timecode will accurately identify every single video frame with a unique number, it isn't precisely measuring REAL TIME.

Say you've edited an infomercial, and began recording it on a piece of timecoded videotape, starting at 01:00:00:00 (called "one hour, straight up").* If the show ends exactly

* Video people never start counting from time zero, but rather start at one hour. This allows machines to pre-roll (back-up) ahead of the first edit, without getting confused.

01:00:00:18
01:00:00:19
01:00:00:20
01:00:00:21
01:00:00:22
01:00:00:23
01:00:00:24
01:00:00:25
01:00:00:26
01:00:00:27
01:00:00:28
01:00:00:29

at $01:29:00:00$ you might be led to believe that your show was precisely 29 minutes long. **THIS IS NOT CORRECT.** Since your videotape is actually playing slightly slower (0.1% slower, to be exact), your actual program duration is almost **two full seconds longer!**

The people who work with videotape often want timecode to do two things: 1) to uniquely identify each frame *and* 2) to give accurate indications of running time. Clearly, regular old timecode doesn't do the latter very well.

REGULAR OLD TIMECODE that has a single number for every frame, that counts from frame :00 to frame :29 and then rolls over — but is temporally inaccurate (by 0.1%) — is called NON-DROP FRAME TIMECODE (NDF) — because it never drops any numbers while it is counting.

The only way to make timecode keep anything close to REAL TIME is to leave out certain numbers. If you skip some numbers (remember that this doesn't affect the video pictures at all; it is only a numbering scheme), your calculations can be extremely close to the actual elapsed time of a segment.

Timecode that skips certain timecode numbers is called DROP FRAME TIMECODE (DF). The way in which it skips is very precise:

Drop the :00 and :01 frame every minute, except for every 10th minute.

01:03:59:25
01:03:59:26
01:03:59:27
01:03:59:28
01:03:59:29
01:04:00:02
01:04:00:03
01:04:00:04
01:04:00:05
01:04:00:06
01:04:00:07
01:04:00:08
01:04:00:09
01:04:00:10

This way, source and record times DO reflect real time, and thus can be used to determine length. (To calculate the length using timecodes, subtract the "in" timecode from the "out" timecode. It can be difficult; you might want to use a special calculator. Editing systems do this automatically.) For many reasons, source material (dailies) tend to be transferred from film to videotape using NON-DROP FRAME TIMECODE — this way, every frame has a number that is one greater than the preceding frame, and the algorithms that convert timecodes into film key (or code) numbers are "safer." In actuality, it makes no difference. Record times for almost all broadcast

television, however, are in DROP FRAME because this way you can easily
see if your program is running long or short — you know how long it will
play on the air. Remember —

> ☞   You *MUST* know whether you are using DF or NDF
>   timecode. It makes a huge difference.
> ☞   *Almost all* nonlinear editing systems deal equally well with
>   Drop or Non-Drop Frame timecode.

FACTOID # 1 • The logic behind Drop and Non-Drop is similar to
what we follow in our calendars for leap years. We pretend that a
year is 365 days long. In reality, a year is 365.24 days long.
Because of this, we drop a day (February 29th) three out of every
four years to "keep in sync." Like with videotape, dropping a day
out of our 366-day year prevents a "cumulative temporal error" in
our calendar. If we didn't correct for it, eventually (in less than a
thousand years), we would be celebrating Christmas in the middle
of the heat waves of summer.
In this sense, our calendar is DROP-FRAME.

WHAT ABOUT AUDIO?

When 1/4" audio tape is synced with film and transferred to videotape
(in a telecine session), it is also being slowed down 0.1% to keep it in sync
with the picture. The speed change is virtually imperceptible, but if you ever
want to resync the audio from a videotape back to the film (either the cut film
or even the original film dailies), it must be sped back up — this is called
resolving.

With nonlinear editing systems, final cuts are created electronically, and
lists for cutting film negative or print can be provided. Assuming your
nonlinear editing system did not change the timings of your edited sequence,
the videotape you have created in offline can often be SYNCED to the film
cut.

But it will not sync directly (remember the videotape runs a tiny bit slow
— even if it is "accurate"). The audio portion of the offline videotape master
is often **"resolved to mag"** to bring it up to the correct speed and record it
onto sprocketed MAGnetic film.

Types of Timecode:
VITC and LTC

There are two ways you can record timecode on a piece of videotape. Running lengthwise along the videotape, as an audio-type signal; or vertically, in each frame, as a video-type signal.

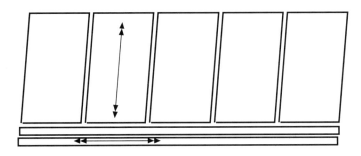

An audio-type encoded signal can be placed on one of the audio channels of the tape (running along the bottom of a tape) or in a special separate channel (running along the top). This timecode, since it runs LONGITUDI-NALLY (*i.e.* lengthwise) along the tape, is called *Longitudinal Timecode* (or LTC). LTC might be called *audio timecode* (if it is recorded in one of the tape's audio tracks), or it might be called *address track timecode* (if it is recorded in the separate address track).

On the good side, LTC is inexpensive to generate, record and read. The encoders and decoders are affordable and are often the choice for budget-conscious facilities and projects.

On the down side, LTC is notoriously prone to getting confused. Because the timecode is running longitudinally, it can only be read by the videotape machine when the tape is moving within a narrow speed range. When you slow it down or stop it, the timecode can't be read. It is very much like a regular audio signal — as you slow down an audio tape, the words on the tape get more and more difficult to understand. When the tape stops, you hear nothing. Because of this, LTC is not the perfect way to edit. The constant slowing down, stopping, changing direction, stopping, jogging some more, then making an edit can cause the machine to "read" timecode a number of frames from where the tape is actually parked.

It is because of LTC (and the now-truly prehistoric *control track* editing that preceded it) that editors first became concerned about editing systems that were FRAME ACCURATE. LTC is not.

If you want really accurate timecode, it must be able to be read while a tape is not "real-time" playing. Here is a video frame — on videotape, and then scanning and refreshing on your monitor:

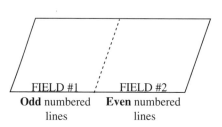

FIELD #1   FIELD #2
Odd numbered **Even** numbered
lines   lines

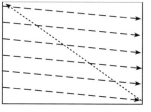

There are 525 discrete lines of video, interlaced (while playing) from two fields on a length of tape. First, Field #1 fills the odd lines on the monitor, then 1/60th of a second later, imperceptible to the eye, Field #2 drops into the even lines. The electron beam in the monitor scans the screen from upper left to lower right, then when it hits the end of the field, seeing no video picture (but rather a "blank" signal), turns off the beam, zips back up to the top and starts again. The first 21 lines make up this blank space between adjacent frames — where *vertical blanking* takes place. Although most

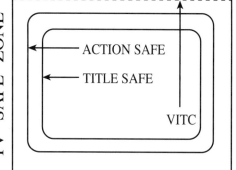

ACTION SAFE

TITLE SAFE

VITC

monitors don't actually show you all 525 lines, you can usually see at least 400 of them (in overscan). Once the picture is on your television at home, you could conceivably see even fewer. This is why there is a TV SAFE zone in the middle of professional monitors that gives you an idea of what TV viewers will see at home.

Since you have more lines of video on videotape than anyone sees, engineers use the blank area for inserting data. Each line has a number, 1 through 525; video engineers often encode the timecode into the picture on line 12 or 14 (or both, just to be safe; lines 16 and 18 are also used). To the untrained eye, a line of timecode in the picture is just a bouncy black and white bar along the top of the screen. But this timecode can be decoded and read by tape machines and, if desired, converted into a human-readable number (by a *timecode reader*) that can be superimposed over the images on a monitor, or recorded into other tapes.

Timecode inserted into each picture frame is called *Vertical Interval Timecode* (or VITC); "vertical" because it resides in what's known as the "vertical interval" between fields. This might appear confusing because although the timecode is roughly vertical on the tape, it is horizontal on the monitor.

In practice, though, VITC is usually used in conjunction with LTC. Being part of the video signal, VITC cannot be read accurately in fast shuttle or rewind speeds, something LTC has no problem with at all.

BLACK

There is a difference between a videotape that is "blank" and one that is "black." A new videotape, just purchased and never before used, is called *raw stock*. If you played it in a VTR, what you would see on your screen would be "static," familiarly called "snow." This is a blank tape. It has no video on it, no timecode, no nothing.

Black is a special video signal. A picture of absolute blackness must be electronically generated and recorded on videotape for you to see, well . . . black . . . on the monitor. When you take a professional videotape with video or whatever on it, and then magnetically erase it (called *de-gaussing*) the tape is returned to a blank state, but not black. A tape used, and then de-gaussed, loses some video quality — slightly more each time it is erased. Professionals rarely re-use a tape for critical applications more than once.

Timecode is recorded, or *striped*, on a piece of videotape to prepare it for use before most editing sessions. In general, black video signal and timecode are recorded at the same time. Remember that just because a tape has timecode recorded on it does not mean that you can see a visible *burn-in* window of that timecode number on the screen. Generating a visible timecode window is an option you have when playing or recording a tape; a window is rarely if ever burned in to black tapes.

Most offline editing, linear or nonlinear, must record onto a *blacked and coded tape*. The computer controllers use that timecode to determine record points, to synchronize the source and record tapes, and to search for specific locations. When these edits take place in offline, they are only recording audio or video over the black, but are not altering the timecode track. This is called *insert editing* (it has nothing to do with the nonlinear "insert.")

If you are editing to all the channels of the videotape (recording to the timecode track *and* control track as well) you are *assemble editing*. Online and offline rarely assemble edit, however telecine sessions often do.

When you assemble edit, you generally record timecode as each edit is made. If this timecode is based on the previous shot's timecode, (continuing it where the previous shot left off) it is called *jam syncing*, and you will end up with a series of edits with continuous timecode running on the tape. A tape made this way is indistinguishable from one that was blacked and coded first, then edited using video/audio inserts; except that after the assemble edit session, the timecode will only extend as far as the last shot. Only **very accurate** timecode generating and editing equipment can assemble edit and pick up exactly where the previous timecode left off.

If you try to assemble edit on a machine that is not capable of jam syncing (since the timecode track will get disrupted at the out-points), the numbers on the tape will not be continuous. When the timecode track is broken in this way, editing can be difficult. You cannot do an insert edit across a break in timecode; computer edit controllers can easily become confused when the timecode jumps for no apparent reason or is missing altogether, and edits and cues may be aborted.

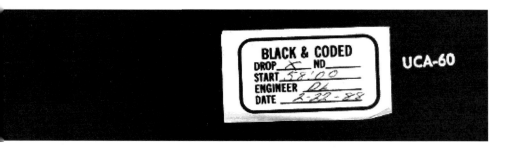

For offline editing, 3/4" videotapes are pre-striped with black and code. **Continuous and ascending** timecode on source tapes is required for the smooth function of many, but not all, nonlinear systems. Systems that continuously monitor the timecode on the tape *can* handle discontinuous timecode, as long as it is unbroken — meaning as long as there are no places where there is NO timecode in the middle of the tape. Editing systems with a log calculating the relationship between the videotape and film must be alerted where discontinuities in timecode affect changes in the relationship to the film. Many systems controlling videotape (whether editing video, digitizing source for a nonlinear session or outputting to tape) will abort recording if it finds a timecode anomaly.

THE EDL
EDIT DECISION LISTS

Prior to 1972, there was no standardized way to repeat the edits of an editing session. Before there was timecode, and before computers kept track of timecode, you could not automatically re-edit old work, or perform an auto-assembly.

CMX pioneered the development of the Edit Decision List, or EDL. Although currently there are other formats for the EDL, in particular those from the Grass Valley Group and Sony, the information contained in the EDL is basically the same.

In the CMX-type of EDL, there are 8 primary columns and 2 ancillary columns. At the top of the page is a header (the title), and below that from left to right, are the following columns: *event number, source reel ID, edit mode, transition type* (sometimes followed by 1 or 2 other columns, *wipe code* and *transition rate*), *source in, source out, record in* and *record out:*

01	02	V	C			02:28:32:03	02:28:49:00	01:00:00:00	01:00:16:27
02	02	A1	C			02:28:32:03	02:28:49:25	01:00:00:00	01:00:17:22
03	02	V	D		45	02:20:37:20	02:20:41:10	01:00:16:27	01:00:20:17
03	02	A1	C			02:20:38:15	02:20:41:10	01:00:17:22	01:00:20:17
04	02	VA1	W	094	60	02:06:00:28	02:06:07:28	01:00:20:17	01:00:27:17
05	BL	VA1	C			00:00:00:00	00:00:01:00	01:00:27:17	01:00:28:17

EDL COMPONENTS

• HEADER: At the head of every EDL is the list's TITLE, and whether or not the record times are in Drop Frame (DF) or Non-Drop Frame (NDF) timecode. Source timecodes are either specified as NDF by colons, or DF by semicolons.

DROP FRAME TIMECODE	01;00;00;00
NON-DROP FRAME TIMECODE	01:00:00:00

Sometimes source timecode-type will vary from tape to tape within an edit session. Some kinds of systems will place a special comment in the list whenever source timecode has switched. For example, it might say "FCM: Drop Frame," meaning that the Frame Code Mode (FCM) had changed, and the new source timecodes are now in Drop Frame.

• The EVENT NUMBER is simply an identifying counter, beginning at 1 and increasing with each edit. While each edit is given its own number, notice that a single edit can take up two lines in the EDL, in the case of dissolves, wipes, and "split edits." The event numer is important in uniquely identifying each edit for eventual list cleaning or assembly — especially since offline editing is usually not done in sequential order.

A **note** in the list is an *unnumbered* comment line, generally associated with the preceding event.

• SOURCE REEL ID is the name of the videotape on which a particular shot was originally located. Every videotape that is recorded in telecine is numbered, and all duplicated reels ("dupes") of that master source tape get identical numbers. For editing, additional copies of a source reel will have identical source reel IDs *with an appended letter*; for example, reel 004 and 004B are identical material. This kind of "B-roll" is usually created for use in effects such as dissolves. In general, but not always, the "hours" portion of the source timecode is equivalent to the reel number. (i.e. videotape reel #4 begins at 04:00:00:00 — four hours straight up.) If the source for an edit is video black, the source ID appears as "BLK," "BL" or sometimes just "L."

Other non-VTR sources, like electronically-generated color bars or other switcher "cross-points" (for doing special video effects) are called auxiliary sources, and are labeled "AUX," "AX" or just plain "X."

• EDIT MODE denotes whether the edit takes place in video only, audio #1 only, audio #2 only, or any combination of these. The letter "B" was originally used to signify a "Both" cut, where both picture and sound cut at the same point, but the introduction of a second channel of audio made this obsolete. "V" is video, "A1" or "1" is audio #1, and "A2" or "2" is audio #2. (Also note that the newer digital VTRs have *four* channels of audio.)

• TRANSITION TYPE describes whether the edit is a "C" for cut, "D" (dissolve), "W" (wipe), or "K" (key). Transitions other than CUTs will be followed by a transition duration, in frames. WIPEs and some kinds of KEYs will have an additional code following the letter, indicating what kind of effect is being triggered. Unfortunately, the wipe codes from the various switcher manufacturers have not been standardized.

• SOURCE IN and SOURCE OUT are the first two columns of timecodes. These numbers describe the timecode of the first and last* frame of the shot, as it is played from the source videotape.

*Actually, the established convention in video editing is that the "out" timecode is that of the frame *following* the last one used (*i.e.* it is not inclusive). This means the next event's **in-point** is the same number as the previous one's **out-point**.

• RECORD IN and RECORD OUT are the last two columns of timecodes. These describe where the source shot is to be recorded on the master videotape. Note that, due to the linear nature of videotape, a change in duration to any but the final shot in a sequence may require ALL following record in and record out points to be shifted earlier or later. Source in and source out points for all these shots do not change. A shift in record times is called a "ripple" because of the ripple effect even a small change makes in all following edits.

All nonlinear systems "ripple" as a matter of design. Oddly enough, *not rippling* is a little less common.

Most electronic nonlinear editing systems identify source material via timecode. Internally the system knows your source in and source out points, and upon command, it can convert that information into an EDL in many types of formats. Some electronic nonlinear systems, however, do not understand timecode: they identify source information in an intermediate format, most commonly videodisc frame numbers. Internally the editing system is still tracking source in and out points, but in disc frame numbers — from these it can derive either timecodes or film edge numbers via some kind of relational database.

The EDL is usually the most important output of a nonlinear editing system. The reason is that these systems are usually used as *OFFLINE* systems, and the creative decisions made in offline must be later conveyed to *ONLINE* for re-creation of the edits using the master source tapes.

FILM EDGE NUMBERS

Running along the side of all 35mm film are some pretty small numbers, updating each foot. In fact, all major types of film have some kind of edge numbering. Edge numbers that are pre-printed on negative stock are also referred to as "key numbers." When edge numbers are stamped onto a positive print of the original negative ("workprint"), and identical numbers are also applied to sprocketed magnetic audio stock to correspond to synchonized tracks, they are called "code numbers," or sometimes "rubber numbers."

Both types of edge numbers are human-readable characters that are essential in the smooth process through post production. Here is a typical edge number on a small piece of film:

The first portion of an edge number is the *prefix*. This set of numbers and letters is constant throughout each original roll of film. For code numbers, the prefix is also used to indicate on which reel a shot originated. Following the prefix is a *footage count*. This number increments every foot throughout the entire roll of film.

These numbers are used to identify any single frame on any roll. Frames without numbers (actually the majority) are identified by counting how far they are located from the closest preceding edge number. This way, every frame can be distinguished by a unique number: the prefix, the footage count, plus the additional number of frames from the location of the actual edge number — for example, MR 2463 +11 is the ID of a frame 11 frames after the key frame MR 2463.

Because the numbers are spaced a foot apart (and there are 16 frames per foot), counting frames from the key number to the frame you want to identify can be tedious and error-prone. Even if you didn't make mistakes, a method of counting still has to be made consistent among the people using the film. You could count from frame 0 to frame 15, then roll over to the next foot (known as zero-counting), but if the next person calls the first frame "1" and counts to frame "16," you'd both be in trouble. Neither is incorrect, but the method used must obviously be standardized. The most common numbering scheme is the 0-to-15 count.

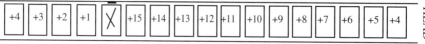

Notice how the head of the film here is reversed, as it would be on a flatbed.

You also must determine from which frame you will start counting. If edge numbers were perfectly placed on the film with relation to the frames, it would be simple to determine the "0" frame (sometimes called a "key frame" or "index frame"). Officially, the key frame is the frame in which the key number *ENDS*. However, in

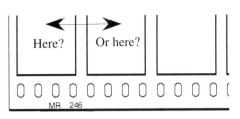

the real world, edge numbers often end on a frame line, making things even more complicated. Everyone using the film *must* determine key frames in the same way.

Tracking key numbers and code numbers is a significant part of the film post production process. Especially with the electronic post of film using video (nonlinear, in particular), the proper entry of these numbers into relational databases is the critical factor that allows film to be conformed from video offlines. The greatest single "frame accuracy" issue in electronic systems has usually been the human error factor involved with the reading and typing of these numbers into computers.

KEYKODE™
MACHINE READABLE FILM EDGE NUMBERS

Kodak has spent a number of years developing a better kind of film edge number for their film stocks. In the early 90s, they introduced KEYKODE™ numbers, which are human- *and machine*-readable edge numbers. The applications of KeyKode have not yet been exhausted. But for electronic nonlinear systems they significantly reduce the human errors involved in the logging of film numbers into relational databases for the eventual reconforming of the negative or workprint.

How KeyKode Works

KeyKode offers a number of significant advantages over the earlier methods of numbering; (remember? a single tiny number each foot of film):

• Along with the alphanumeric edge numbers are machine-readable bar codes. A bar code scanner can easily read and input edge numbers into computer logs and databases without the necessity of human reading and typing.

• Edge numbers have also been made easier to read by eye, spaced better, and have an additional digit added to the prefix, which minimizes the possibility of duplicate prefixes on separate rolls of film.

• A special dot has been inscribed following each number, denoting the Zero-Frame. This should standardize the reading of edge numbers, and eliminate discrepancies created from different counting schemes.

• An added mid-foot edge number has been included, printed smaller than the regular footage count, and should decrease problems associated with identifying short (less than one-foot) shots.

• There are a number of marks, checks and symbols located on the stock that make verification of frame lines, negative matching, and offsets much simpler. There are also film stock identification codes, manufacturer identification codes, and various product and emulsion codes.

KEYKODE

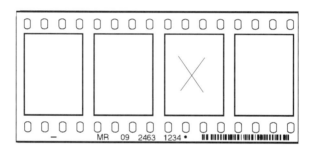

KeyKode is rapidly being standardized as an industry-wide change in film numbering. A number of the major film manufacturers have already adopted its use, and ancillary products are available in many film and video post production settings. Bar code readers can now be included in telecines, in film bays, and anywhere that the reading of edge numbers is done. In particular, telecines fitted with KeyKode readers can begin the logging process automatically during telecine. Accuracy of editing systems, speed in telecine and efficiency in the preparation to edit are all significantly increased with the adoption of this technology.

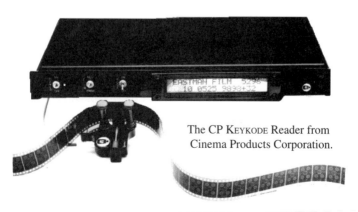

photo by Joel Lipton

The CP KEYKODE Reader from Cinema Products Corporation.

T E L E C I N E
GETTING THE FILM INTO THE VIDEO WORLD

Here is a piece of film, 35mm, 4 perf, shown actual size:

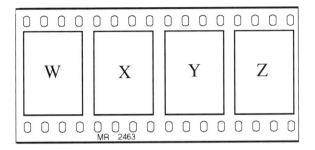

As you may recall, most film is played at **24 frames per second**. Another way of saying *24 frames per second* is that each single frame is 1/24th of a second long.

If a single film frame is 1/24 of a second long, then two frames are 2/24ths (1/12) and three frames are 3/24ths (1/8), and so on:

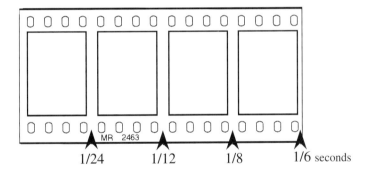

If you continue with this, you can see that the time it takes 4 film frames to pass is 1/6th of a second. Exactly. Remember this for later.

AUDIO, unlike picture, does not have "frames" and thus does not move at 24fps — it just keeps on moving all the time, smoothly and constantly. While your picture jumps from frame to frame, the audio (usually on 1/4" magnetic tape) plays smoothly along.

VIDEOTAPE is also different from film. It doesn't look the same. It isn't
recorded the same. It doesn't play the same. One-inch (1") videotape —
although one continuous strip — is usually diagrammed something like this:

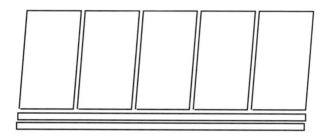

Notice that each videotape frame is slanted. This is because, unlike 1/4"
audio tape, videotape runs across the play/record heads of a VTR at an angle
(actually, it's wrapped around a cylindrical head, but ignore that right now):

AUDIO   VIDEO

Also, you may remember that videotape records and plays at **30 frames
per second**, so that if we do the time thing again we would see:

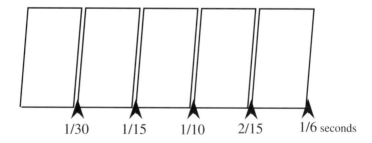

1/30   1/15   1/10   2/15   1/6 seconds

— where the first frame is presented for 1/30th of a second, two frames play
in 2/30ths (1/5)... and so on. Unlike film, on video 5 frames play in 1/6th
of a second. And if this wasn't strange enough, every single video frame,
although uniquely identified by a timecode number, is ACTUALLY made

up of two nearly identical **VIDEO FIELDS** (known quaintly as "field 1" and "field 2"). So let's do the videotape diagram again, this time with fields drawn in, and timecode labeling each frame:

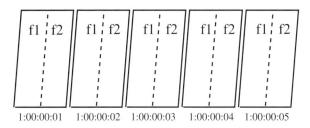

    1:00:00:01   1:00:00:02   1:00:00:03   1:00:00:04   1:00:00:05

Every frame has its own timecode: either encoded longitudinally, in an audio track on the tape (called "Longitudinal Timecode," or LTC), or encoded in the picture portion of the tape, vertically in the space ("interval") between field 1 and field 2 of each frame (called "Vertical Interval Timecode," or VITC). Or in both.

When videotape is recorded — for instance, when you go out with your VHS camcorder — the same image is being recorded into both field 1 and field 2 of every frame.

FACTOID #2 • The image in field 1 is only half the video image — but not the top half, nor the left half, but the *odd* half. A video image is made up of **525** lines of video per frame; field 1 contains the ODD numbered lines (1, 3, 5 ... 525) while field 2 contains the EVEN numbered lines (2, 4, 6 ... 524).

When the tape is playing, each full video frame is the result of interweaving (called "**interlacing**") of the two fields. When you inspect a freeze-frame on videotape, you are only seeing <u>one</u> of the two fields — or HALF the resolution of the moving image.

A TELECINE is a machine that transfers FILM to VIDEOTAPE. No simple task. You put in your film (either positive print or original negative), load up a record videotape (in a format of your choosing), and get to it. But how? Film runs at 24fps and videotape records at 30fps.

If you transfer (copy) one film frame into each video frame, look what happens:

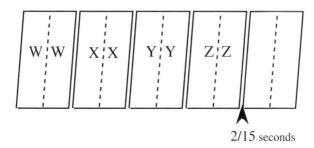

2/15 seconds

We took **4 FILM FRAMES** — that should go by in 4/24th of a second, and copied them into **4 VIDEO FRAMES** — which play by in 4/30ths of a second — FASTER than film! In fact, this kind of direct transfer will speed up the action on film by **20%** and, as you can imagine, it throws the audio on the 1/4" tape completely out of sync with the picture.

This kind of transfer is called a **2:2 transfer**, because every film frame is copied into 2 video fields, repeatedly (they could have called it a 2:2:2:2:2:2 . . . transfer, because that's exactly what's happening).

If you want your videotape copy to play at the same speed as the original film, you have to take every FOUR film frames and *SOMEHOW* stretch them into FIVE video frames.

A telecine transfer does this by performing what's called a **3:2 transfer** (or **"pulldown"**). Rather than copy each film frame twice — once into each of the two video fields — a 3:2 pulldown transfer will make every other film frame *a little bit longer* . . . it will copy every other frame into one extra field:

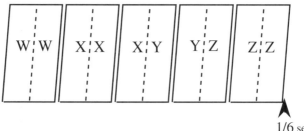

1/6 second

The key here is that it is okay for certain video frames to be made up of two fields with *totally different pictures* in them. Technically speaking, some film frames have been transferred to 1/30th of a second-types of video frames (2 fields long), and others have been transferred to 1/20th of a second-types of video frames (3 fields long). This is a 3:2 pulldown (3:2:3:2:3:2....)

Now 4 *film frames* occupy the space of 5 *video frames* — all playing in 1/6th of a second — faithfully remaining in sync with the 1/4" audio tape.

To make matters slightly more confusing, it turns out that there are actually FOUR different "ways" that the 3:2 transfer can take place:

- Starting on field 1, alternating 2 fields, then 3...2...3...2...3...
- Starting on field 1, alternating 3 fields, then 2...3...2...3...2...
- Starting on field 2, alternating 2 fields, then 3...2...3...2...3...
- Starting on field 2, alternating 3 fields, then 2...3...2...3...2...

They ALL produce the same end result. So what's the difference between them? In some ways, nothing, at least when you are editing videotape. BUT, if you ever want to convert your videotape (timecode) edit decisions back to FILM, the kind of 3:2 pulldown is **extremely important.**

On every videotape created from a 3:2 pulldown, regardless of the KIND of 3:2 pulldown that was done, the 4 transferred film frames are converted to one of FOUR TYPES. See if you can follow this:

TYPE A: a 2-field "frame" beginning on a field 1
TYPE B: a 3-field "frame" beginning on a field 1
TYPE C: a 2-field "frame" beginning on a field 2
TYPE D: a 3-field "frame" beginning on a field 2

To demonstrate this, let's look at all four of the possible 3:2 pulldowns. Below each transferred film frame is its "3:2 frame-type":

1 Starting on field 1, alternating 2 fields, then 3...2...3...2...3...

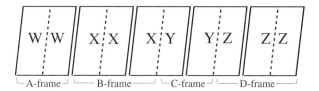

2   Starting on field 1, alternating 3 fields, then 2...3...2...3...2...

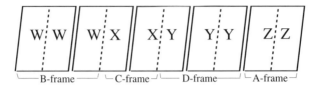

3   Starting on field 2, alternating 2 fields, then 3...2...3...2...3...

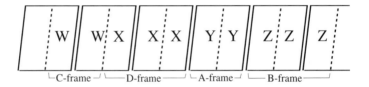

4   Starting on field 2, alternating 3 fields, then 2...3...2...3...2...

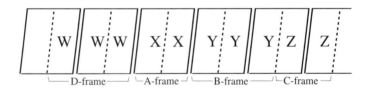

It should be clear from these diagrams that no matter what type of 3:2 pulldown is performed, only four *kinds* of transfer frames are created.

All editing systems that deal with both film and tape will *NEED TO KNOW THE EXACT RELATIONSHIP BETWEEN THE ORIGINAL FILM AND THE VIDEOTAPE*. This relationship is always entered into some kind of database (or **"log"**) as an "interlock" point between film key numbers (or code numbers) and a videotape timecode. In addition, the 3:2 type IS REQUIRED.

Film frame "W" for example, may have originated as key number MR 2462+15. On videotape, after a telecine of type 2, it will have a timecode reference of 1:00:00:01, and will be a **B-frame**. If, however, it is a telecine of type 1, the same frame will have the same references, but will be an **A-frame**.

CUTTING FILM USING VIDEOTAPE
FRAME ACCURACY

Any piece of film can be transferred to videotape for editing with little difficulty; 3:2 pulldowns are standard operating procedure for all telecine bays — that's exactly what a telecine machine is designed to do.

Once your film is on videotape, you do not need to go through all these frame and field gymnastics, if you don't want. Film is edited on videotape all the time. Linearly. You can always take a 3/4" tape with a burned-in window of timecode and offline to your heart's content: the online of this edit session can air on television, can be dubbed to other tapes for distribution, whatever.

The issues associated with cutting film on tape do not pertain to projects finishing **only** on videotape. As long as you never want to go *BACK* to having cut film, there are no concerns of "frame accuracy," or "slipping sync with audio," or any of those things that everyone talks about when discussing electronic film editing.

But, to edit on videotape or some video media and then to *CUT NEGATIVE OR POSITIVE PRINT* — that is where the 3:2 pulldown and issues of frame accuracy come into play.

Let's say you take a 3/4" dub out of telecine: you've transferred your film and now want to edit on tape (linear or nonlinear, it makes no difference). Here's your videotape:

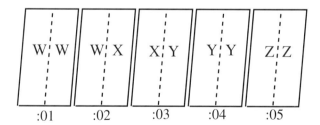

You can see the 3:2 pulldown frames clearly here. Every film frame transferred to this tape can be classified either an A, B, C or D-type frame. This transfer is of type 2, as the first frame is a B-frame (a three-field frame beginning on field one). Remember, it doesn't matter what kind of transfer you do as long as you know what kind of 3:2 frame you *began* with.

Now, let's say you decide to cut between tape frame :01 and :02:

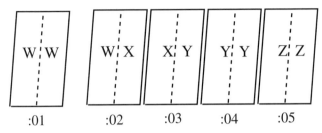

The deceptively simple question is, "Where is this location on the film? Is it between frame W and frame X?" The answer is . . . sort of. It certainly looks like it could be. Realize that *all* videotape, when parked in "freeze frame" can only show you ONE FIELD. Which field? Either one, it depends where you stop. So when you park your videotape, and choose to edit AFTER frame :01 (in this case, in the middle of an A-frame), you still have one more field of film frame W after this point . . . hmmm . . . okay . . . so maybe cutting your film between frame W and X isn't the right spot. But what else could it be? Cutting between :01 and :02, and cutting between W and X both give you a 1-frame shot . . .

Okay, we'll skip this one and try making an edit on our videotape between :02 and :03.

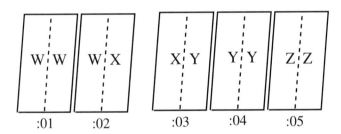

Not better . . . in fact, considerably worse. Now where should you cut your film to match this edit decision? Video frames :02 and :03 have split film frame X! "When I made my decision to cut out on video frame :02, did I mean to *INCLUDE* or *EXCLUDE* the image in film frame X?"

Since all videotape only shows you one field at a time, we might be able to say "If you were looking at field #1 of both frame :02 and frame :03 when you made the decision to edit here, then you *were* deciding to cut between W and X, *NOT* X and Y." But with videotape editing, the EDL will not take into account which field you were looking at when you made your editing decision.

To make matters smoother, when regular offline editing systems DIGITIZE video they save space by only digitizing each FIELD #1 (yes, this gives you ½ the resolution of the image). So unlike interleaved video, if you present each field #1 from the videotape, the editor would see this:

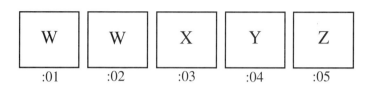

DIGITIZED VIDEO

The 5 video frames look like 4 film frames, with the W frame printed twice. Now, it is pretty easy to determine visually where to cut the film. *So what that you have two W frames*, it still has a direct visual relationship to the film (and, no, you can't edit between the two W's). Note: Some digital systems can also digitize at 24fps — meaning that they just ignore the repeated field. This saves disk space and reduces motion artifacts, as well as eliminating the need for any 3:2 calculations when calculating the film cut.

Videodiscs have another way of handling the 3:2 sequence. Although all videodiscs are dubbed from a videotape, and all videodiscs *physically* PLAY at 30fps, they all can be encoded with electronic marks, called *WHITE FLAGS*, which allow the disc to show you only *real* film frames.

These white flags signal appropriate machinery that a NEW film frame is beginning on a certain video field. They would be placed like this:

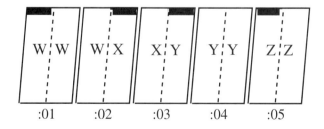

WHITE FLAGS

If the person making the disc takes the time to "white flag" it, disc players can "read" these white flags and show the viewer clean and unique film frames. This way, every frame an editor moves through on a disc has a single one-to-one relationship with the original film. *Visually*, this is an excellent way to accurately edit electronically, and can be performed by discs as well as many digital film-cutting systems that digitize at 30fps.

But there are more than VISUAL issues involved with accurately cutting film from video media.

There are also very significant issues of TIMING.

For example, if I am editing a very special "flash cut" kind of sequence for a musical section of a film, and I make 24 one-frame edits — I have made 24 extremely careful edit decisions. I have chosen 24 special frames and timed them to the music.

But on my electronic editing system, each of the 24 frames, even if *each* one represents an *actual* and easily identifiable film frame, is still playing on my video media — where each frame plays at $\frac{1}{30}$th of a second. That means that this sequence appears to play in $\frac{24}{30}$th of a second.

But when I go to cut my film, and cut all 24 of these frames, it will take an entire second to play ($\frac{24}{24}$ths of a second). I won't be in sync with my film music track. (Actually, **no** electronic system can be temporally accurate if *all* edits are 1-frame long. Some of the edited shots need to have at least 2 video frames in order to be adjustable.)

Many electronic editing systems will automatically alter the "film cut" to time out to the duration of the video cut. These systems use information about the 3:2 sequence to calculate film-cutting locations that will allow pictures to remain in sync with audio. *Even if you have carefully chosen* each of the 24 frames, by the time this sequence appears on film, it may only be 19 or 20 frames long in order to have the same duration.

The question arises "What is frame accuracy?" Is it an exact frame-for-frame correspondence between the video editing media and the final film? Or is it a temporal exactness that makes 100 minutes on video become exactly 100 minutes when the final film is conformed?

There is no one answer. Electronic editing systems deal with these issues slightly differently.

It may sound harsh, but in general, editing is a forgiving art. Even the most exacting and precise editor, when cutting narrative material, would not notice if you added or dropped a frame here and there in his or her cut to make it run the "correct" length. If you removed a single frame randomly from a project, the odds are good no editor would notice. Only in very tight scenes (or extremely fast cut sequences, like the example above) might anyone notice something missing. Yes, once in a while an edit is precise and a frame added or subtracted would change everything, but this is the exception and not the rule.

All of these things are based on the nature of the 3:2 pulldown.

The simple fact is that virtually ALL nonlinear editing systems are editing timecode, just like traditional videotape editing systems. They create timecode lists, and then 'trace' those numbers back through a relational database to generate the film numbers. It is simply because there are 30 frames per second in video that there is *no* one-to-one relationship between the editing media numbers and the original film.

This is not always the case.

The only way to have visual *and* temporal frame accuracy is to actually have a one-to-one relationship between the film and video media. One way to do this is to play and edit your videotape at 24fps — like the film. If you have special tape decks and an appropriate edit controller, you can do a 2:2 telecine transfer and edit *frame accurately*.

Another way is to use certain videodiscs. Although videodiscs physically play at 30fps, it is possible to encode the frame numbering so that there are not 30 frame numbers per second, but rather 24. The good news is that any kind of videodisc player can read the "white flags" and *show* you "real" film frames. The bad news is that only *unformatted* discs can encode disc frame numbers in this unique 24-frame way. These discs have both advantages (accuracy for film) and disadvantages (cost, availability), but regardless, they are a good way to avoid the issues associated with timecode conversion to film numbers. *Remember that the accuracy is not just a function of the editing media, but rather HOW an editing system DEALS with numbers in that media.*

Digital media is perhaps the best choice, as it can digitize video at either 30fps or 24fps, *if it has appropriate digitizing hardware and software.* Not all digital systems can digitize only 24fps; more common is a digitizing process that records 30 frames each second, but only identifies 24 with unique numbers (exactly like a laserdisc with white flags). This latter method provides identical accuracy to a real 24fps digitization, but still requires more space and contains motion artifacts.

DETERMINING FRAME ACCURACY TYPE

If you want to ascertain, for whatever reason, whether your system is editing "film frames" and then calculating video; or "video frames" and calculating film, there are a number of simple tests you can perform. These will give you some idea as to the scope and degree of accuracy a system possesses.

QUESTION 1: Cut a list of short edits. How long is your video EDL? How long is your film cut list? Now, add a SINGLE frame to the head of the very first shot. Print out a new EDL and a new film cut list. What changes? A one- or two-frame modification of a single edit on a 30fps-based system will result in a change in many film edit points throughout an edited sequence.

QUESTION 2: Will an auto-assembly (either the online session or a simple 3/4" output from the nonlinear edit session) play in sync with the cut film? Are they the same length? (Don't forget that you still must resolve the video speed from 29.97fps to 30fps to make them sync.)

When most systems claim frame accuracy, they are not lying. Because videotape is NOT film, and because of the 30fps of video, there is no identical location to cut film when editing tape. All video systems try to make the best possible choices in the cutting of film — best visually and best temporally.

For many theatrical films, cutting negative directly from the nonlinear system is not an option; there will ALWAYS be a workprint cut. Because of this, any problems that might have been created can be easily fixed on film. Many editing systems allow for cutting negative DIRECTLY from the final cut on the editing system. Productions can utilize these features to save the huge costs associated with printing the negative. Most nonlinear systems will generate a negative cut list, and will edit for temporal frame accuracy — keeping the picture in sync with the sound. Because of the forgiving nature of editing, the changes in the cut made by timecode-based calculations are usually imperceptible.

But the only way to have both visual and temporal frame accuracy is to have the electronic system based on the same frame rate as your final product. Although *all* systems must use a mathematical algorithm *somewhere* in the video-film conversion, it is much more *forgiving* to convert film numbers (at 24fps) into a frame accurate EDL, than it is to convert timecodes (at 30fps) into a frame accurate film list. This is solely due to the fact that the smallest unit of film (and therefore the smallest error) is 1 frame = $\frac{1}{24}$th of a second; for video it is smaller: 1 field = $\frac{1}{60}$th of a second.

PREPARATION FOR TELECINE
"THE FILM FINISH"

Telecine is simply the transfer of film — negative or positive — to videotape. For projects that are finishing on videotape, there are no special precautions to take. But for projects finishing on film (or *both* film and tape), there are some important specifications that must be met in the telecine. It is *THIS* process that will ultimately allow the film to be cut correctly.

Like printing dailies on film, the telecine session usually involves choosing selected takes to transfer to tape, not entire rolls. Every time you choose another take, you must stop the telecine machine, find where you want the take to begin on the film and the videotape, sync up the 1/4" audio, and then record.

Because of the need to sync the audio tape, you are stopping and starting recordings on every single selected take, even if they are already in the right order and consecutive.

To use this tape in editing electronically, the 3:2 sequence (3:2:3:2:3:2 ... *etc.*) must be maintained through the entire tape, with no mistakes. This means that if the last take ended on a A-frame, the next take recorded MUST begin on a B-frame:

Scene 9D, take 24 ⟶ ⟵ Scene 10A, take 2

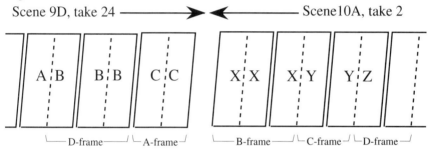

If this relationship is changed, the editing system may not able to determine the accurate position of the film relative to the videotape. There are a number of ways to make this process go smoothly in telecine:

1. Assemble negative or positive into entire reels of the exact length needed for the tape — then telecine the entire roll at once. If you don't stop the telecine recording, the 3:2 sequence cannot be accidently changed. Later, go back and record the sync audio from the 1/4" tapes.

2. Standardize telecine edits. Make every edit go in on an A-frame, for example. This means that the film edge numbers are not consecutive and MUST be logged as a new entry in the editing system's relational database. Often, this is of little consequence since transferring the select takes from the negative involves skipping whole areas of film, and the film numbers are already going to be discontinuous.

When telecine edits are standardized, it is easy to know that when timecode :00 is the beginning of an A-frame, every timecode multiple of 5 frames is also going to be an A-frame: :00, :05, :10, :15, :20, :25, :00. . .

AA	BB	BC	CD	DD	AA	BB	BC	CD	DD	AA	BB	BC	CD	DD
:00	:01	:02	:03	:04	:05	:06	:07	:08	:09	:10	:11	:12	:13	:14

3. The newest method (but one which may not be available at every facility yet) is to use KeyKode™ stock with appropriate readers. This automates the process, and can generate a machine-readable EDL-of-sorts, indicating the key number, timecode, and 3-2 type (among other data) at every edit. If this data list is converted to a format that an editing system can read — like OSC/R or Evertz —jumps in 3-2 type can be managed.

Without the advantages of KeyKode, and short of using one of the other two methods, the telecine operator or assistant must check the tape at each edit to determine the 3:2 frame type for the last recorded frame. They would do this by looking for white flags, or by actually viewing each frame of tape, field by field.

Another piece of equipment helpful for the telecine of material finishing on film is the Time Logic Controller, or TLC. This is a box that accurately counts film frames in telecine. This is one way it is used:

1. At the head of each film reel loaded onto the telecine machine, a single frame is scribed or punched to make it easily recognizable via picture. The punch is made at a key frame — a frame with a clear film edge number. This number is entered into the TLC.

Now, wherever you roll in your film, the TLC will show the film edge number of that frame on its LED display. This number can be noted or encoded into VITC on the videotape for retrieval later.

2. At each telecine edit, even when one or more film frames are going to be skipped in the transfer, the TLC continues to track the film frame numbers. The TLC displays film numbers accurately, regardless of what is being transferred.

3. At the end of the source film reel, the last key number is punched and checked against the TLC readout: they should be identical. This provides a good double-check for the accuracy of the film numbers for each edit; it confirms that there are no misprints of the film's edge numbers, and no mistakes in the videotape's recording.

If the last punched key number matches the TLC, you can be insured that ALL points in between are accurately encoded.

WHO CARES?
THE REAL LIMITATIONS OF "FRAME ACCURACY"

Although it seems reasonable to be concerned over the visual accuracy of electronic film systems, in actuality, this may not be as important as most people would believe. For cutting film, there is really no way to be visually accurate on a 30fps-based system. So although people speak of visual accuracy, they tend to be more aware of *temporal* accuracy. As long as edited sequences are durationally accurate, pictures will sync to audio tracks built on film or tape and will play fine on television. And since no one will really notice the difference between an electronic cut and its film cut, should we really care about this type of frame accuracy?

It is a valid point. Most productions that are cutting film are doing one of two things. If they are television-based, the film images will be as identical to the cut images as the media will allow; and they sync up to real time. So visual "inaccuracies" are irrelevant.

If the project is theatrical, like a feature film finishing on negative, serious productions are going to conform the lists to workprint before cutting negative. If there are any problems with the cut, and they would only be minor, they can be corrected on film. The video system is then only a tool to create a first cut, and future changes will be done on film. This kind of project is also undaunted by visual-accuracy issues.

Neither of these situations implies that the errors are not present. It only means that in the real world of editing, the errors are of little practical concern to the people using the equipment.

In instanceswhere absolute frame accuracy is imperative: where work-print changes are being done on the system, or where negative is being cut directly from the system, only 24fps coding — found with some digital systems can achieve the essential 1:1 relationship between the editing media and the film. **True 24fps digitizing is the best choice** from both a *visual* and *temporal* accuracy standpoint; still, it is only in the rarest of situations where issues of frame accuracy at these degrees find themselves in the forefront and where some of the differences between these methods are of concern.

THE BIG QUESTIONS
"WHY DOES THE U.S. HAVE 24FPS FILM AND 30FPS VIDEO?"
"WHY IS PAL 25FPS?"

For years, longer than there has been video, film has been recorded and projected at 24fps — in all nations. When individual frames are shown in succession 15 or more times per second, they begin to fuse together and give the perception of motion. Clearly, the faster you display frames, the smoother the illusion. For a variety of perceptual and economic reasons, modern film makers decided upon 24fps as a good speed.

When video was pioneered, it too was intended to play at around 24fps. However, at the time, you needed some kind of clock (or constant pulse) to synchronize frames to play at a constant speed. The easiest location of a natural synchronizing ("sync") signal was the alternating current (A.C.) in an electric socket. In the US, A.C. power is 60Hz. This provided the basis for our video sync. Although video has 30 frames per second, it is really 60 fields per second — 60 *anything* per second is 60Hz. A perfect source for sync for the early black and white televisions.

In Europe, as in many nations, A.C. power is at 50Hz. This would mean that a 25fps video signal is more natural (50 fields per second). It also means that film productions shot at 24fps could be transferred to videotape at 25fps and the change would be so small that no one would notice (a 24fps to 25fps speed-up is about a 4% change in run time; 24fps to 30fps is a 20% increase).

"WHY DOESN'T NTSC VIDEO *REALLY* RUN AT 30FPS?"

It used to. The first black and white televisions, getting their sync from wall current, really played at 30.00 fps. The problems arose when Americans wanted to add color to the black and white signal. The addition of a chrominance signal to the luminance signal required the NTSC people to come up with a very slight modification in the sync signal (and thus the frame rate). To explain exactly why adding color to black and white necessitates this change would require a detailed explanation of video signal encoding and wave physics. That isn't going to happen here.

Anyway, with the development of color TV, the perfect 30fps signal was changed to the irregular 29.97 signal — a change of 0.1% — but requiring the subsequent development of such things as drop-frame timecode and sync generators. We still refer to NTSC video as 30fps video because "30" is a more convenient number and is sufficient for most discussions.

COMPILATION REELS
THE BUILDING OF "LOADS" FOR EDITING

Productions are hardly ever shot in scene order. Films are shot in an order that maximizes efficiency in using locations and scheduling actors. Because of the non-order of the shoot, film rolls tend to have scenes on them somewhat randomly. And this can cause some trouble for nonlinear editing systems.

Clearly, for some production-logical reason, if scenes 43 and 75-76 were shot around the same time, they may be together on a single 400 foot roll. In film post production, the circled takes will be printed and delivered to the film assistant for syncing and coding. Then the rolls are broken down by scenes and takes for the editor's convenience.

In electronic post, the negative is developed and sent with the 1/4" sound rolls to telecine. Dailies material is transferred from film to videotape reels, and synced in the process.

But nonlinear systems must have all the dailies associated with a particular scene accessible to the system at one time. Regardless of when a scene was shot, all the affiliated shots need to be grouped.

All systems have physical or economic limitations as to how much material can be accessed simultaneously. For some systems it is only a few hours of material. For others, it could be 4 hours. (Digital systems configured for feature films often store more than 50 hours.) Many systems have diminishing performance as the quantity of source material increases; digital systems always trade off *load quantity* for *image quality*.

Regardless, **when the total quantity of source material is greater than can be fit on the system at one time**, the material must be broken down into "chunks" that can be handled at once, and that will be edited together. These chunks are called *loads*.

---

deluxe laboratories
1377 North Serrano Ave., Hollywood, CA 90027 (213) 462-6171

CAMERA REPORT
SOUND REPORT   No. 23701

DATE _8 - 15 -91_   CUSTOMER ORDER NUMBER ____
COMPANY _OTM_
DIRECTOR _GRANT_   CAMERAMAN/RECORDIST _JOAO_
PRODUCTION NUMBER OR TITLE _RUST_
MAGAZINE NUMBER _2107_   ROLL NUMBER _A 125_
TYPE OF FILM / EMULSION _5296-634-192_
PRINT CIRCLED TAKES ONLY: ☑ ONE LITE   ☐ TIMED

SCENE NO.	TAKE	DIAL	PRINT	REMARKS
75 D	①	5	㉚	
	②	10	㉚	
E	1	17	70	
	③	27	⑩⓪	
76	1	35	80	
				G 200
				NG 150
				W 50
				SE 0
				T 400
400'				

For the older tape-based systems, a load is generally the quantity of material that can fit on a videotape — usually around 3 hours. The usual way to break down source material is into groups based on the final cut's "acts." It is as though you have four boxes for four acts. Every time you get a shot that goes in Act 1, you set it in the Act 1 box. Likewise for material for Acts 2, 3, and 4. This way, material that goes together is loaded together on a single set of tapes. This is called "building compilation reels," or *comp reels*.

Compilation reels are built in one of two places:

1) In the telecine session
2) On the editing system

Most commonly, the 3/4" simultaneous dubs of source material are delivered from telecine to the editing bay. These tapes contain the source footage in "daily" order, exactly how it was shot and found on each camera roll. Then, using the editing system (usually) the shots on the tape are logged, sorted by scene number, and then copied selectively onto the editing media.

On videotape-based systems, this process of building "loads" is central to the systems' design (all duplicate tapes are built simultaneously). In many cases, loads cannot be mixed, and thus absolutely all material needed for an act must fit on the appropriate tape. On digital systems, selected scenes are digitized from the videotape onto specific magneto-optical (MO) disks or hard disks. On pre-formatted videodisc systems, daily tapes are either dubbed at one time to discs (as if they were tape-based systems) or dailies are very selectively dubbed as necessary — thus saving valuable disc space.

Less common and more expensive, compilation reels can be built in telecine — this involves the transfer of original film negative to various videotapes as neccesary to divide up the source material into organized pieces. This kind of compilation is generally only done for the pre-mastering of unformatted (ODC) laserdiscs — as they can not be added to, and their capacity is under 30 minutes.

Regardless of whether this work is done in telecine or on the editing system, it is generally done or supervised by the editorial assistant.

The key to compilation of material is to insure that all source material for a single scene (or set of scenes) is accessible for editing at the same time.

On some systems, considerable work must be done to insure that material is loaded together. On projects where there is a large amount of source footage relative to on-line hard disk storage,especially when shot over a long span of time or when scenes are shot incompletely, compilation is a necessary organizational process for nonlinear systems.

For shorter projects, like music videos, industrials, and commericials, generally no compilation is done. For these (or any other projects) to have full nonlinearity, all source materal must fit into a single load. Once a record tape is used in conjunction with editing, virtually any size project can be cut regardless of load size, although some degree of nonlinearity is sacrificed. Comp reels are very common in the editing of feature films, movies-for-television, and some episodic TV programs, as all are shot over many weeks, often using hundreds of thousands of feet of film.

PREPARATION FOR COMP REELS

Part of the job of the Assistant Editor is to organize the source material for editing. The first question that must be answered is this: "How much footage is expected for this project?"

The next question is: "What is the size and nature of the source material on the editing system?" Based on the answers to these questions, compilation can begin.

For example, if a project expects to have 10 hours of selected takes for a television program, and you happen to know the program is divided into four acts, you might be able to break the dailies into four loads of roughly 2.5 hours each or two loads of 5 hours each.

Now, let's say you are using an editing system that is configured to handle a maximum load of three hours at your chosen image resolution. No problem. You will divide all your dailies up into one of the four acts, load each scene onto one of the four *act* "drives" (or set of drives) and you even have some room for a little slop — some acts might be a little shorter, some a little longer, but your margin for error is pretty good.

Better yet, let's say your system holds a 6-hour load. You can't fit the entire pile of dailies in the system at once, but you only have to break down the footage by pairs of acts: Acts 1 and 2 in one load, and Acts 3 and 4 in the other. Much easier. And much more forgiving (say, if a scene is moved from Act 1 to Act 2).

Now, what if your project involves 30 hours of dailies? It is a feature film. Like most films, it can be somewhat "naturally" broken down into 10 or so "reels." So on this project, you might divide dailies up by approximate reels — by sequence, perhaps. This kind of project can require a great deal more effort in organization, especially if loads cannot be mixed on your nonlinear editing system.

On projects that involve extensive organization (or projects much largers than the editing system's source capacity) — regardless of whether storage is on hard disks, MO disks, videodiscs or tape — the following method often proves helpful:

Build a chart with scene numbers across the top — beginning with Scene 1 and ending with the last scripted scene. Beneath each scene number will be a column for shoot days. The shoot days should be filled in using the one-line notes generally prepared before a shoot and held by the 2nd AD (or perhaps the director). *Someone* always knows roughly when each scene of a film will be shot. Even if the days end up wrong, they will generally give you an idea of approximate shoot date and *relative* time. Here is an example of such a chart, beginning to be filled out:

SCENE	1	2	3	4	5	6	7	8	9	10	11	12	13	14	15	16	17	18	19	20	21	22	23	2
DAY	16	16	5	5	5	12	1	18	18	18	18	31 33	31 33	31 33	16 33	6	6	6	6	6	2	2	2	1
Script Page	1		2			4		5				6	10				11					12		
LOAD																								

Next, depending on how much work you feel like doing, you can read or skim the script to find "sequences" that, although they have different scene numbers, generally must edit together. These sequences should be the smallest editable 'chunks' of the project. Note the sequences on the chart. Usually, sequences are shot around the same time. You also might include the script pages where scenes begin.

Now, from looking at this chart, it is easy to locate problem areas.

For complex productions, these kinds of charts take a lot of the guess-work out of building compilation reels. They prevent the accidental situation where new material won't fit in a load, and thus a scene or sequence cannot be easily edited electronically. As the production continues, and projections become fact, the chart can be updated — adding information about scene footages and load groupings.

TELECINE LOGS

A combination of developments in the 90s have made it possible to automate many parts of the logging procedure for nonlinear system projects finishing on film. The core of the new automation is due to Keykode™ — machine readable negative key numbers. These numbers are bar codes that run along the film that can be read by a barcode reader.

By employing a barcode reader, and taking the numeric output of the reader into other telecine devices, every edit made in telecine can be recorded, including the exact key number where each shot begins, the exact timecode of the tape frame where the shot is being recorded, and the 3:2 telecine frame type.

Prior to the evolution of these automated functions, telecine assistants were required to physically punch the negative at the head, read off the key number at the punch, whereby the telecine operator hand-typed the key number into a counting device (like a TLC), which displayed the *calculated* key number at any given frame. During the telecine edit session, a paper list was generated that showed the key numbers at each edit, along with the timecode. This list was taken to the nonlinear editing system along with the 3/4" videotape dubs made in the session, and hand-typed into the log of the nonlinear system. Regardless of how "accurate" the editing system was, there were numerous points in the process that allowed for human error: reading the key number in telecine, the telecine operator typing it into the system, the assistant mis-reading or mis-typing the number from the paper list into the nonlinear editing log. . .

With Keykode™, the key numbers are automatically entered into and moved through the telecine process. In telecine, the edits are recorded along with any additional information the telecine operator or assistant chooses to type (usually the scene and take numbers, and sometimes a brief description of the shot). As well as managing all this data, the **telecine "log"** that is being created also includes the audio timecodes from the 1/4" Nagra tape that is being synced to the negative (these numbers will be important after the editing, during the audio conforming or layback).

The telecine log is a special kind of *Edit Decision List* (EDL), but on this new kind of EDL the source column contains film key numbers, and the record column contains timecodes. There is not yet a standardized format for the telecine log, and so the manufacturers of some telecine equipment have developed their own. There are three principal telecine log formats:

Evertz offers **KeyLog™**
Adelaide Film Works offers **OSC/R™**
TLC offers **Flex™**

```
110 Scene  41A       Take  2        Cam Roll A83       Sound 24          17:55:46:22.0
200 IMAGE  35 23.98 KJ455232 005255+10 000070+14 Key EASTM KJ455232 005255+10 P1
300 VTR-A ASSEMBLE  1211      At 11:07:54:20.0 For 00:00:47:05.0 Using VITC AUTO
300 VTR-B ASSEMBLE  VC19      At 00:07:54:15.0 For 00:00:47:05.0 Using TIMER 1
400 SOUND 29.97 fps 24        At 17:55:46:22.0 For 07:03:24:08.0 Using LTC
500 GV100  1  Cut             to 24            Fx        Rate       Delay
```

An example of a Flex™ format printout

In an attempt to standardize these varied formats, the SMPTE has also organized a committee to develop a standardized SMPTE Telecine Log format, although it has not yet been completed.

Regardless, these telecine logs, like EDLs, can be output directly onto floppy disks. Then the disks can be taken to any of a number of nonlinear editing systems and input directly into to the system (although it should be noted that not every system can read every format of telecine log). In this way, human error has been minimized, and the accuracy of the editing systems for film cutting has been effectively increased.

Log Date: 07/13/93	LAW AND ORDER						Page 1 of 2	
EPISODE TITLE	69009			ACT	COMP REEL	901		
EPS #		MEDIA #						
Scene	Take	Time Code	Fr	Cam	Prefix	Key	+Fr	SNDR	Nagra TC
16	5	01:00:00:00	A	A1	KJ15 5796	5184	+15	1	09:58:36:11
16	6	01:01:06:25	A	A1	KJ15 5796	5289	+07	1	10:01:36:04
16A	2	01:02:17:05	A	A1	KJ15 5796	5452	+13	1	10:13:06:28
WT1002	1	01:10:09:05	A					1	12:53:27:10
9	1	01:10:37:05	A	A5	KJ06 5688	4965	+08	2	15:45:22:18

An example of one facility's proprietary telecine log format printout

COMPONENT AND COMPOSITE VIDEO

PART I

Before we discuss component and composite video, let's talk about television history. When TV first came out, it was showing images in black and white. The technical term for a video signal that is in black and white (*i.e.* monochrome) is a **luminance** signal.

A decade later, scientists figured out that they could broadcast and record in *color*. Problem was this: a lot of Americans already had black and white sets — and they weren't just going to toss those out. Our government decided that whatever kind of color signal was broadcast, the old black and white sets still had to display the images. So rather than design a really great color set with an ideal color video signal, the scientists had to compromise. Since black and white TVs could only read luminance broadcast signals, color signals had to be part luminance and part . . . something else. What they added (for everyone with color sets) was a **chrominance** signal to go with the luminance. A color set could combine both signals together and produce a color TV picture; an old set would just ignore the chrominance.

Thus, today we have inherited an electronic video image that is made up of these two kinds of signals: **luminance** and **chrominance**. Luminance is the brightness of the signal, from black, through grey, to white. Chrominance (or "chroma") is the color part of the signal, relating to the hue and saturation. You could think of it this way: luminance draws a picture and chrominance paints it.

PART II

If you recall fingerpainting in kindergarten, you might remember that you could create any color in the rainbow by mixing three primary colors. The "additive" primary colors, as they are called, are red, blue and yellow.

Light is a little different than fingerpaint. If you mix colors of light, they don't turn black (like the paints), but rather they turn white. For light, the primary colors are red, green, and blue.

PART III

The smallest unit of a video picture is called a *pixel*. You can see the pixels if you push your face close enough to a TV screen. For a color TV, every phosphor dot is illuminated by one of three electron guns inside the set, and aimed (from behind) at the screen. One gun shoots at red phosphor dots; one at green dots; one at blue. It is through the careful mixing of the amounts of red, green and blue that you get a pixel of any other color.

Since you can create virtually ANY color from red, green and blue, it follows that you can also create all the shades of grey — from black to white. Sound familiar? That's what luminance is. This means that a luminance signal can be created from red, green, and blue; by definition, so can chrominance.

If you have the time (and money), you can record and play the red, green, and blue signals separately (but simultaneously). That's what a component video signal is. You are keeping the three colors as separate *components*.

COMPONENT VIDEO is a kind of video signal where the RGB color information, either in analog or digital format, is maintained in separate channels, thus using separate cables and separate internal processing for each color.

This causes two things. On the good side, the red, green, and blue signals stay very pure and very clean. Your pictures look good and suffer little generation loss with each dub. On the bad side, it is expensive — you need special equipment that maintains the separate-ness of the colors. You need at least three cables of exactly the right length for the video signals. To use component video, you really need component tapes, component VCRs, component switchers. . . .

COMPONENT

COMPOSITE VIDEO is much, much simpler. Composite video takes the red, green and blue signals it "sees" and mixes them together into a single signal. Get it? It *composites* them. The method by which the RGB signal is compressed and encoded together is based on government-approved standards (like NTSC, PAL, or SECAM); in the US we use the NTSC standardized specifications for encoding the RGB signals — this is known as NTSC video.

TV receivers decode (and decompress) the NTSC video to play on monitors. On the good side, composite video is inexpensive and simple. The signal may be marginally more mushy, but it doesn't become a problem (or even noticeable), until you have re-recorded the signal many times. This encoding and decoding degrades the quality of the colors, and thus the accuracy of the chrominance and luminance information. Almost all professional equipment is composite.

Component video, because of the huge additional expense and trouble, is really only necessary where multiple generations of a videotape are necessary; for example, in special effects.

Encoder   Decoder

COMPOSITE

Actually, there is a third option. If you don't want the trouble and expense of three cables (RGB), and don't want to mush everything together in one cable, you can use *two* cables*. If you have two separate cables, you can separate the luminance (or "Y") in one cable, from the chrominance (or "C") in the other. This is what high-end consumer video does in the form of S-VHS and Hi-8. They are not quite component, but not as mixed-up as composite. Technically, they are called Y/C video (or *pseudo*-component):

Encoder   Decoder

Y-C VIDEO

All broadcasts in all nations are composite video. The *method* by which the video is composited, however (NTSC, PAL, *etc.*), varies by nation.

Internally, cameras produce (and monitors require) red, green and blue. However, it's unwieldy to feed RGB signals from the multitude of sources through all the intermediate steps that lead to broadcast (*i.e.* recording, editing, effects... up to transmission and reception). So pictures are encoded into composite video at the camera and left that way until they get to your TV. Besides, the less they're decoded and re-encoded, the better they look.

---

*In practice, these "two" cables are generally combined into a single multiple-wire component cable with a standard S-video connector.

DO NOT READ THIS UNLESS YOU ARE REALLY SMART
OR REALLY CARE

To understand how luminance and chrominance can be related to red, green, and blue, first you need to know a little something about colors.
Lesson one:

$$Red + Green + Blue = White$$

This means that if you mix the three primary colors you will have perfectly white light.
Lesson two:
Scientists enjoy making 3-dimensional graphs for discussing colors — with red on one axis, green on another, and blue on the third. Every color imaginable is on this graph, somewhere, with three coordinates — a value of red, green and blue.

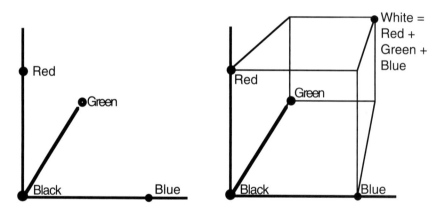

If you have trouble viewing the drawing in 3-dimensions, try visualizing the corners of a box.

In this box, black is at the zero-point for all axes. The corner opposite black — the corner where you are adding red, green and blue — is white.

Since there is a "corner" that is black and a "corner" that is white, we can easily draw a line between them. (See the figure at the top of the next page.) This line denotes a perfect gray scale (all the shades of grey from black to white). This line is the *luminance*. For no obvious reason, scientists call luminance "**Y**."

Chrominance, in one manner of viewing things, is everything else.

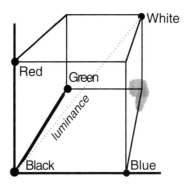

Let's review. The bottom left corner is black. One adjacent corner is red, one is green and one is blue. The far corner is white. But what are the other three corners? Answer:

$$Red + Blue = Magenta$$
$$Blue + Green = Cyan$$
$$Red + Green = Yellow$$

There are electronic scopes that analyze video signals. These scopes have rotated the RGB cube so you are looking down the "barrel," so to speak, of the luminance. Black and white are in line.

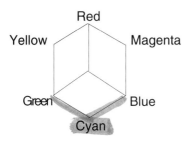

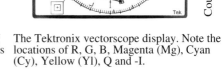

Imagine the black corner is in the center of the diagram, and the "line" towards white is running directly away from you.

The Tektronix vectorscope display. Note the locations of R, G, B, Magenta (Mg), Cyan (Cy), Yellow (Yl), Q and -I.

Now that we are talking about rotating the RGB cube, it's worth noting that scientists tend NOT to view the cube as an RGB cube. Like on the scope, they make the luminance (Y) one axis, and then they create two other axes at right angles to the luminance, called, in the US, "I" and "Q." YIQ components make up NTSC video signals. A more international standard (as in PAL broadcasting) uses "U" and "V" to create YUV "color space."

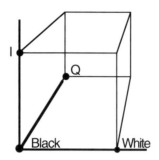

Chrominance = I and Q

"RGB" is used by engineers for television and computer CRT (*cathode ray tube*) monitors, and "YIQ" or "YUV" when working with broadcast signals. Artists find all of this too technical; they describe their color space as being made of HSB (hue, saturation, and brightness).

So, some video technicians talk about "RGB" — but what they are really talking about is usually "YUV." In some sense, there is really no difference.

To make matters slightly more complicated, our RGB cube is not particularly *cubic* in shape. If you added equal amounts of red, green and blue, and showed it on a color TV, you still wouldn't get white (as you would expect from reading this). In the real world of CRT monitors, the color formula for white is:

30% Red + 59% Green + 11% Blue = White

This formula is also how engineers derive luminance:

30% Red + 59% Green + 11% Blue = Luminance (Y)

So, although the reality of this "cube" is a little distorted, the fact remains that luminance and chrominance broadcast signals can be created from the colors red, green and blue. To record video in a camera, or play video on a monitor, you must manipulate RGB. To broadcast, you must convert the RGB to NTSC composite video. But to record and edit, you have choices: either component or composite can be used, each with its own costs and benefits.

COMPONENT & COMPOSITE VIDEO SYSTEMS

You can't always tell from the name of a video product whether it is component, composite, or Y/C video. All video, even digital video, must be encoded as one of these types. Below is a list of common video formats:

COMPONENT:
D1 (19mm digital)
D5 (19mm digital)
BetaSP (1/2" analog)
MII (1/2" analog)
Betacam (1/2" analog)
Digital Betacam (1/2" digital)
DCT (19mm digital)
(4:2:2 = digital component)
(RGB = analog component)

Y/C (or *pseudo-component*):
S-VHS (1/2" analog)
Hi-8 (8mm analog)

COMPOSITE:
D2 (19mm digital)
D3 (19mm digital)
1" type C (1" analog)
3/4" U-matic (3/4" analog)
3/4" SP (3/4" analog)
8mm (8mm analog)
VHS (1/2" analog)
(NTSC = analog composite)

Although D1, D2 and D3 tapes are all 19mm formats, the actual tape used is different and the cassettes are not interchangeable. The same is true for the 1/2" formats.

When editing and manipulating video, it is the encoding and decoding between component and composite that can degrade the signal. Unfortunately, due to the costs of component video, many facilities use some component equipment, like D1 tape machines, and interface them through *composite* switchers. Realize that you will only receive the full benefits of component video when the *entire* video path is maintained as component.

19mm is approximately 3/4" in width; 8mm is approximately 1/3"

VIDEOTAPE QUALITY

It is worth noting that although the Y/C formats theoretically could produce a better signal than the composite formats, the tape stock used for consumer videotape is of significantly lower magnetic oxide densities than those used for professional purposes. That oxide density, when combined with tape speed, defines the maximum *signal bandwidth* that can be recorded.

Also note that tape **width** (1/2", 3/4", 1") has no direct correlation to videotape quality. The first videotapes were 2" wide, and they are virtually extinct today. Only the area of tape covered by the record heads with each scan (for each field) combined with the density of oxide particles in that area, affects video quality.

Oxide particles on videotape are in some ways like sand on sandpaper. Course sandpaper has few grains per inch, like inexpensive tapes. Very fine sand on high grade paper has thousands of grains per inch, like high density and professional tapes. The more density, the more detail.

1/2" tape

When the tape is moving slowly, the video fields are not very long, and have small area.

...but when the tape is moving quickly, the same video field is longer — and has more *area*. More area means more oxide per field, which means better looking images.

1" tape

Larger tape formats might mean better images, but the real factor is the area. The fields on a 1" tape are about the same size as the faster moving 1/2" tape in the middle figure.

FACTOID #3 • NTSC video is well known to have 525 scan lines (or "rows") per frame. However, the horizontal *detail* in those lines (or "columns") varies depending on the recording format. This approximate value is often refered to as "horizontal resolution".

Laserdiscs .... 425-450 columns
S-VHS tape .... 400 columns
VHS tape .... 200 columns

CABLE CONNECTORS

Cables are relatively new territory to film people, but common to the video and computer literate. Perhaps you have played with your home cable TV hookup — then you've seen *coaxial* video cable with a threaded connector; stereophiles may have seen *RCA* cable plugs; every Walkman has an *RCA-mini* headphone jack. These are all just different kinds of cable connectors. It would behoove anyone who works with video and computer equipment to be familiar with the names of the most common connectors. Here, for the uninitiated, is a quick review:

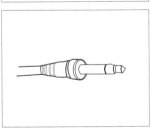

BNC

Convenient because of its twist-lock connector, BNC is a common type of video connector. It is found on most professional coaxial cables — for the video and timecode -in and -out of professional VTRs. Component video systems often use bundles of 4 video cables with BNC plugs — one each for red, green, blue, and sync signals.

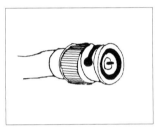

RCA

These very common connectors can carry either video *or* audio. They are found on most consumer video and audio products. A signal travelling through this cable is "unbalanced"— *i.e.* somewhat prone to picking up extraneous noise and hum.

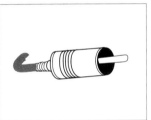

RCA-mini

These compact connectors carry audio. They are found on most consumer audio products in the form of headphone plugs and jacks. Occasionally, VTRs and monitors will have RCA-mini jacks for audio-in and -out.

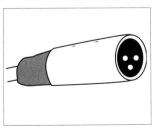

XLR

These large 3-prong connectors are used for audio in most professional equipment, including VTRs. Unlike cables with RCA plugs, XLR connectors generally carry "balanced" audio signals. The connectors have a small latch that snaps into the jack to hold them in place; the release switch is on the top of the jack.

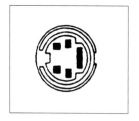

S-Video
This consumer video cable is for the two-channel (Y-C) video signal used in S-VHS and Hi-8 VCRs and cameras. (The connector looks identical to the Macintosh ADB* connectors — the mouse and keyboard plugs).

Computers have different types of cables. They don't carry video and audio, just digital data. Computers move data in one of two ways: either in serial or in parallel. The most common method is serial. There are two common protocols (like languages) for transmitting digital information serially. They are called **RS-232** and **RS-422**.

Most computer peripherals use the RS-232 protocol, most commonly in editing systems for computer-to-peripheral machine communication. RS-232 cables have varying sizes and shapes of connectors. In whatever kind of connector is used are a number of "pins," (tiny metal wire endings) which transmit certain information. RS-232 connectors often have 9-pin or 25-pin configurations (called D-9 or Din 9; D-25 or Din 25). These connectors are usually "D" shaped to prevent them from being connected upside down. They tend to have snaps or screws on either side to prevent them from accidently being moved once attached.

Computer cables, like all cables (for digital or analog video and audio), have certain length restrictions. Certain computer cables cannot run longer than 50 feet. Video cables can need signal amplifiers if their lengths are too extended.

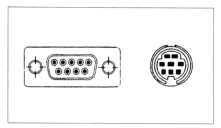 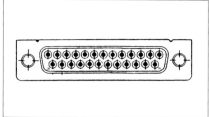

RS-232 connectors
D-9 and mini circular-8 are on the left, D-25 is on the right.

*"ADB" stands for Apple Desktop Bus, the proprietary Apple connector style.

VIDEODISCS
PART ONE

In 1958, an inventor named David Paul Gregg coined the word "videodisk" by combining the word "video" from videotape and the word "disk" from the computer business. The invention was to be the recording of video by frequency modulation (FM) on a optical medium. Later, the "disk" was changed to "disc" to specify a consumer application. Today, they are called many things: *laser discs*, or *video discs*, or *video disks*. To be most correct, the generic term for the video media is "videodisc," with the term "laserdisc" specifically being the trademarked name of all LaserVision format videodiscs. Although there is no formal rule, "DISK" with a "K" usually refers to computer disks, and "DISC" with a "C" refers to videodiscs.

Introduced in the early 1970s by two companies independently, MCA in the U.S. and Philips in the Netherlands, videodiscs could have been subject to the format standardization risks manufacturers find when inventing new products. To avoid needless problems, both companies decided to get together to standardize a format for recording the discs . . . and did so by 1974. They called it LaserVision.

Although other formats have come and gone, the LaserVision format, now owned by Pioneer (they bought it in 1989), is the only consumer format for videodiscs. More recently, Pioneer changed the official format name from LaserVision to LaserDisc.

All videodiscs are ANALOG video recordings. Although they share many features with Compact Discs (CDs), they are not digital as many people believe. They can be encoded with digital audio alongside the analog video, but some video quality is sacrificed when CD-quality audio is placed on a disc.

They are played via small, low-powered lasers located in the bottom of most players — either infrared diode lasers or red gas (helium-neon or "he-ne") lasers are most common. So far, no available videodiscs are erasable.

Videodiscs can be made in one of two ways: they can be recorded one at a time, from a videotape source (like 1" or D2 or 3/4"), much like any video dub, or they can be recorded onto a master disc and then stamped out in mass quantities, in the same way LP records are stamped out in plastic.

When making fewer than 10 copies of a disc, the one-at-a-time method, called "one-off" duplication, is the most economical. For mass market discs, like movies for resale or interactive entertainment discs, the master disc pressing method is optimal. All disc-making as it relates to nonlinear editing and all test or check discs for multimedia are "one-off."

There are primarily two kinds of videodiscs: *unformatted* and *pre-formatted*. Unformatted videodiscs (called *Recordable Laser Videodiscs* or RLVs) are like raw unexposed film stock — there are no frames predetermined on the film; the camera creates the frames and frame numbers during the exposure. When you unload a partially-used roll of film from a camera, it is impossible to re-load it later and start taking pictures right where you left off. You must estimate your stop-position, and re-start past it. But the odds of starting exactly where you left off, or on an even multiple of a frame boundary, are slim.

Unformatted discs are the same way — it is extremely difficult to stop recording on them and then start again. For a number of reasons, including the inability to reframe a video image, you cannot add to an unformatted disc for editing.

Pre-formatted discs can be used a little at a time. You can start recording on any given frame number because all the frame boundaries and all the frame numbers are already on the 'blank' discs. You need but record the video/audio you desire in the empty spaces.

Because of existing laser technology, a 12" laserdisc has exactly 54,000 tracks that can fit on it — either preformatted or unformatted. At 30fps, with one frame per track, this translates to 30 minutes of video.

On preformatted discs, each one of these 54,000 tracks has a predefined disc frame (track) number. However on unformatted discs, this number can be placed and used selectively. Because of the 3:2 pulldown done in film-to-tape telecine, many video frames actually consist of two separate film frames. All videotape (and videodiscs) can be encoded with special signals called "*WHITE FLAGS*" that indicate just where each new film frame begins. This can allow all disc players to display only CLEAN and unique film frames when paused. If tapes are not white-flagged, or if the disc is dubbed without using the flags, you will see "flicker frames," where both fields of each video frame are presented — two out of every five video frames is made up of fields with different images. These images will appear to stutter or flicker on a monitor.

Unformatted RLV discs, however, can adapt disc frame numbering to point to each unique film frame, using the white flags. In this case, though, a disc that still holds 30 minutes of program material will only have 43,200 individual disc frames (instead of 54,000). Although visually identical to preformatted discs, numbering white flags rather than disc tracks allows computers to relate each disc-frame with a one-to-one relationship with the original film from telecine, if this is desired. Pre-formatted discs must identify frames with one of 30 frame locations per second.

Unformatted discs must be recorded on a special and expensive disc recording machine made by the Optical Disc Corporation — the only company that currently makes equipment for this process. The machine is large (about the size of a washing machine) and costly; thus it is relatively uncommon in the post production facility business. However, these ODC RLV discs are the only recordable videodiscs that use the LaserVision format,

An ODC Recording Station — the disc machine is on the left, 1" source is on the right. *Photo courtesy of ODC.*

hence RLV laserdiscs can play in any consumer disc player. Blank unformatted discs cost around $40 each.

Blank pre-formatted discs are somewhat more expensive, around $150, but the recorders for these discs are small and relatively inexpensive (under $30,000), making them extremely desirable for many editing situations and some animation layoff. Unfortunately, there is no preformatted disc available that records in the LaserVision format. Only RLV discs offer this format.

VIDEODISCS
PART TWO

All 12" videodiscs, whether pre-formatted, unformatted, or just purchased at Tower Video on Sunset Boulevard, can have a maximum of 54,000 tracks on a single side. Like an LP record, discs have one long spiraling groove, beginning in the center and spinning out toward the outside edge. However, this single groove BEHAVES as if it were actually a set of loops of concentric circles, where the first loop is at the center of the disc and lies nested in the next loop, and so on, moving toward the outside. For our explanation, we will treat discs in this way. In other words, one track = one loop.

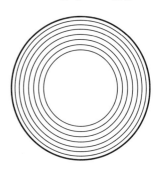

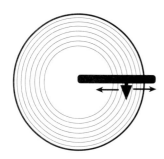

The laser rests in the bottom of a videodisc player, aiming upward at the disc. The disc starts spinning the moment it loads into the player, and doesn't stop until you remove it. Imagine a videodisc player as an upside down record player. But unlike a traditional record player, 1) the head never touches the disc, and 2) the head is on a fixed armature that can only move radially — in and out — over the disc. To cue to a desired location, the armature need only locate the exact radius of the track where the frame is located.

Because of this, a disc can be played very quickly or very slowly with great ease. To play fast or slow, the videodisc itself doesn't need to speed up or slow down, only the laser riding on the armature. In fact, to play a freeze-frame you need only stop the laser in a single location. The disc itself will continue to spin. But since the laser is stopped, and only scanning a single loop, you see a single frame.

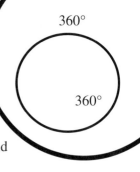

If you think about the diagrams above, you might realize, "Hey, the loop closest to the center of the disc is a lot smaller than the loop that is on the outside edge." You also might say, "If every loop is a single frame (as has been implied), why is the frame in the center so much smaller than the frame on the outside? Don't you have a lot of wasted space on the outside ring?"

This would be a good observation. To be technical for a moment, a circle is 360°. All circles are made up of 360° — big circles, small circles, whatever. The length of a circle (the linear distance around a circle) is called the CIRCUMFERENCE. Frame 1 on a disc and frame 54,000 have different circumferences, although both frames have 360°.

If you unwrapped both frames, and made a straight line out of each circle, it would look something like this:

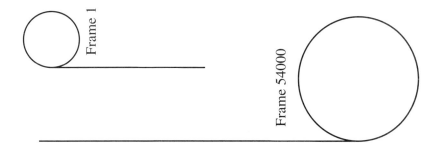

Clearly, frame 1 is shorter in "length" than frame 54,000. If the loop way out on the edge of the disc is so big (in reality, about 36 inches), and frames can be as small as frame 1 appears to be (about 14 inches), why don't we just pack MORE frames in every loop? If you can unwrap a frame and make a line 14 inches long, then you should be able to fit a few frames in the outside ring. It wouldn't be one frame = one loop (360°), it would be one frame = 14 linear inches. This way you could fit more on the disc.

In fact, you can do this. There are two kinds of discs: those with one loop equal to one frame, and those where all the frames are the same linear size and are packed in wherever they fit. *To play correctly, a videodisc player must play all frames at the same rate, wherever the frame is located on the disc.* If you use the kind of disc where each loop is a single frame, the only thing constant about all frames on the disc is the ANGLE that sweeps out each frame, in other words, 360°. All frames are 360°. To play this kind of disc, you must spin it at a **Constant Angular Velocity** or CAV.

However, a disc that has frames packed in as tightly as they will fit, has many frames per loop. The angle of a frame changes, depending on where the frame is found on the disc. Frames at the edge of the disc have smaller angles and frames in the center take up larger angles. The only thing constant about this kind of disc is the LINEAR size of each frame. Whether on the

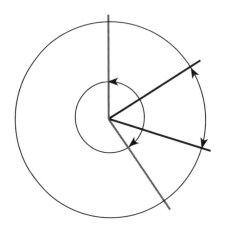

edge or in the middle, each frame is the same length. These discs must play with a **Constant Linear Velocity** or CLV.

A Constant Angular Velocity (CAV) disc spins at a constant speed (*rpm*). Like an LP record. A Constant Linear Velocity (CLV) disc must speed up or slow down, depending on where the laser is on the disc; it must turn slower at the edge, and faster in the middle. (Note: this has no bearing on how fast the images on the disc *appear* to be moving). All laserdisc players will play either CLV or CAV discs. Players automatically adjust depending on what kind of disc you place in them.

A videodisc player cues by moving the laser head to a single loop (or track) on the disc. If you only have one frame in that loop, you can access any frame of the video material you desire by cueing to its unique disc frame number. But if you want lots of material on the disc (and thus have to squeeze in more video frames) a single loop might have a handful of frames there. How can you pick just one? How can you freeze-frame?

Fact is, you can't. CLV discs do not allow you to freeze-frame, or play in slow motion, or cue to individual frame locations. In short, you can't edit using a CLV disc. On the other hand, you can pack a full hour of material on a CLV disc. Although it's no good for editing, it is okay for watching films. CLV discs that have commercial movies on them are often called "Extended Play" discs. If you look at these, you can see this noted on the covers.

CAV discs, although they only hold 1/2-hour, are in many ways excellent for analog video editing. They are perfectly frame accurate, and since they show you both video fields at once, the resolution is twice that of looking at a frame on videotape (which only can show you one field at a time).

Videodiscs can play in extremely slow and high speeds (20X) with negligible picture breakup, and if modified, videodisc players can play back audio without pitch change as you speed up or slow down. This is an exceptional feature of videodiscs.

D I G I T A L   A U D I O   A N D   V I D E O
A Basic Introduction

This is a wave of audio:

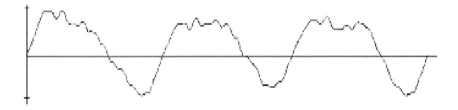

In many ways, it is just a more complex version of a simple sine wave, like this:

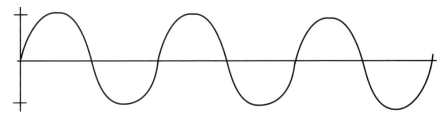

Audio waves, like ocean waves, go up and down through time. Exactly how quickly they go up and down over time is defined as the wave's **FREQUENCY**.

A high-frequency wave might look like this:

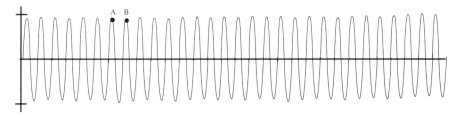

As you can imagine, if you were listening to this wave it would be a high note — the frequency of a wave is commonly called its **PITCH**. On graphs like these, the passing time is measured along the horizontal line (or X-axis, if you're into this stuff), beginning at *time zero* on the left and continuing toward the right.

A low frequency note (perhaps made by a tuba), looks calmer:

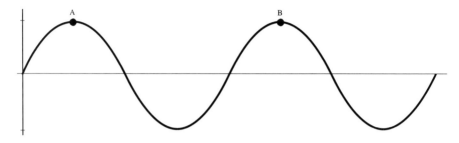

A wave is a series of repeating curves. One chunk of the wave, the smallest amount that repeats, is called one cycle, or one period:

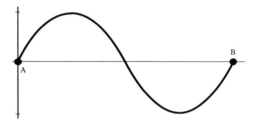

The time it takes for exactly one cycle — or the distance between any two corresponding points of consecutive cycles — is called the **WAVELENGTH**. On these waves, two points have been marked (A and B). The distance between these two points is each wave's characteristic wavelength. If you examine the diagrams, you can see that a high frequency note has a *shorter* wavelength; a low frequency note has a *longer* wavelength.

Don't go on if this isn't clear.

The frequency of a note can be measured by its wavelength. You might say that a high note (with high frequency) does this: ⌒⌄ a lot — many, many times a second. The wavelength of this high note, let's say, is 1/1000 of a second. Another way of looking at it is that the wave can do this: ⌒⌄ (one complete cycle) 1000 times per second.

THE NUMBER OF TIMES A WAVE MAKES A COMPLETE CYCLE in one second is measured in units called "Hertz" and is abbreviated "Hz." Our high frequency sound, then, could be described as a 1000 Hz tone; our low frequency sound is a 10 Hz tone.

FACTOID #4 • Your ears (and brain) have the ability to perceive sounds with frequencies as low as 10 Hz and as high as 22,000 Hz. Sounds higher or lower than that range are inaudible. The range from 10 to 22,000 Hz is the human ear's FREQUENCY RESPONSE.

Let's go back to our low frequency (10Hz) note:

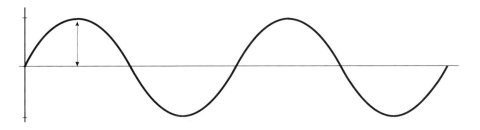

The amplitude of the wave (the distance from the "mean" or "zero" line to the wave) defines a wave's *energy*. In general, this energy is perceived as **VOLUME**. (For technical reasons not worth going into, this energy is measured in volts.) The amplitude of the wave, then, corresponds to how LOUD the note is.

Both the wave above and the wave below depict a 10Hz frequency:

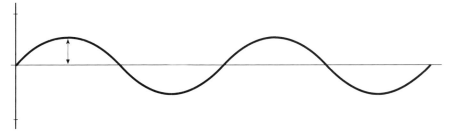

The only difference is the amplitude. This second example is a "quieter" note of 10Hz, and the top example is pretty loud, but both are the same note: that is, they both have the same frequency.

In technical terms, a wave is continuously changing energy (amplitude) over time. If I asked you how loud this note was, you might have to ask me "when," because as you can see, the height of the line changes constantly.

To **DIGITIZE** an audio wave, you must be able to describe it with numbers — with coordinates, like any graph. (*Remember that the x-axis or horizontal line is time, and the y-axis or vertical line is intensity or "volume".)

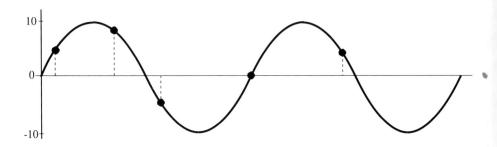

Let's choose a few places to check to see what the coordinates are: Now, we have selected 5 points in our wave, and by writing down the time and the "volume" at each of the times, we have created a small table:

TIME (x) (in seconds)	VOLUME (y) (in volts)
0.02	7
0.035	8
0.059	-5
0.1	0
0.145	4

And if you replotted these 5 points on a new, clean graph, it would look like this:

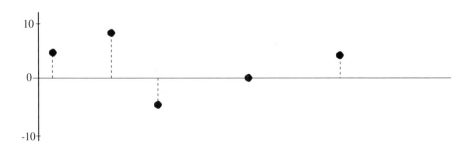

But many waves could be drawn through these points, like this one:

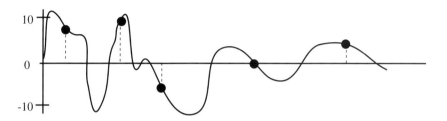

In the first place, taking a wave and converting it into *numbers* is called **ANALOG to DIGITAL CONVERSION** (often abbreviated "A-D" or A-to-D conversion). Taking numbers and re-generating an analog sound is D-A conversion. In this example, we first went A-to-D, then we went D-to-A. But as we can see, this wave is NOT identical to the 10 Hz wave we started with. What happened? As you may have figured, when you digitize a wave, you must check the amplitude (volume) of the wave *VERY OFTEN* . . . and for a variety of reasons, you want to do it at a constant rate:

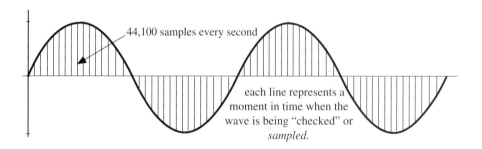

44,100 samples every second

each line represents a moment in time when the wave is being "checked" or *sampled.*

The more often you check, the more data in your table, and the more accurate your recreation of the wave will be. HOW OFTEN YOU CHECK the amplitude is called your **SAMPLING RATE**.

You don't want to sample TOO often, however, because at some point you will be taking more samples than you need; realize that a computer doing all this sampling has to remember every point you sample, and more remembering means more storage needed. For sampling to re-create those extremely high frequency notes at the upper end of your hearing range, scientists (in particular, one guy — Joseph Nyquist) determined that you must sample the wave at least *twice* as fast as the "fastest" frequency (or highest note).

Since we can hear notes up to around 22,000 Hz, the necessary rate of sampling for digital audio is usually greater than 44,000 Hz. This means that the computer (specifically, the A-D converter) is checking the analog music 44 thousand times each second to store the value of the wave at that moment. Compact Discs (CDs) for example, sample at 44.1 KHz.

As with any graph, you must determine just how precisely you want to measure the volume (power) at a given point in time; you do not have an infinite number of amplitudes on the vertical axis. If you start at "0" and the maximum power at the top of the y-axis is "10," how many volumes are in between? Ten? Fifty? One hundred?

The number of values on the y-axis determines how precise your data will

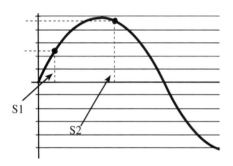

be — the more values, the more precise. You want to be accurate *ENOUGH,* but like sampling rate, there are costs in being *too* precise.

Since all computer counting is done using binary numbers, you have the following choices for how many values you want running up the y-axis: 2, 4, 8, 16, 32, 64, 128, 256, 512, 1024 . . . and so on. The same kinds of scientists who decided that 44.1 KHz is sufficient sampling decided that 65536 values was enough accuracy for re-creating extremely high quality audio. "65536," is 2 to the 16th power. In base two (called bi-

At both of these sampled points (S1 and S2), the values on the y-axis fall between two measured levels; the reconstructed audio wave would therefore have some error. By adding points to the y-axis, accuracy is increased.

nary mathematics), this means that for every sample taken, a 16-bit number is remembered (some value between 0 and 65535). CDs have 16-bit audio. If you want to save space, you could sample less often (*e.g.* 22KHz, 11KHz) or you could drop your accuracy from 16 bits to something like 8 or 4-bit audio. Dropping the sampling rate will lose some higher frequen-

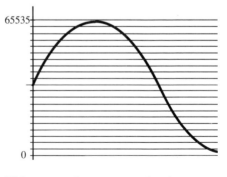

cies; dropping the number of bits will increase the amount of noise.

DIGITAL VIDEO

VIDEO can also be digitized, but un-like audio, it is much more complex. But why? A moving picture must be sampled in both *time* and *space*.

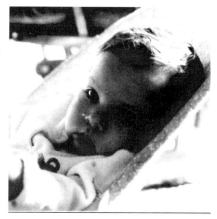

Sampling in time is what a film camera does — then the projector shows different *still* images presented in succession 24 times each second. Somewhere around 10 or 15 frames per second, a set of discrete images will begin to fuse together (animate) and you will begin to perceive continuous motion.

Sampling in space is what captures each single image. For film, it is the microscopic grains of silver halide in the emulsion — each crystal responds uniquely to light. For video, it is a screen of dots. A digitized image looks a lot like a TV image. The pictures must be broken down into tiny pieces, small enough that the viewer only sees the big picture and not the discrete pieces.

A sample of video, the smallest "chunk" of a video image, is called a PIXEL (short for *Picture Element*). If you look closely at your TV or computer monitor, you can see them. Technically speaking, a home television set has relatively big chunks — a fairly small sampling rate. Although home televisions display about 400 horizontal lines on a screen, the master videotape consists of 525 lines per screen; and each line contains about 640 pixels. Thus, for each frame of broadcast video, you have about 336,000 pixels of information (640 x 525 = 336,000). Actually, only 480 of the 525 lines are used for video picture, so the real number equals 307,200 pixels (640 x 480).

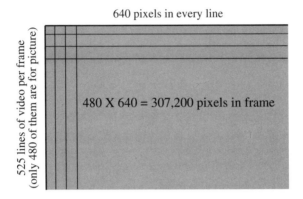

640 pixels in every line

480 X 640 = 307,200 pixels in frame

525 lines of video per frame
(only 480 of them are for picture)

But what is the information? In our music analogy, it will be amplitude (loudness) and frequency (pitch), that change in time. For video, there are also two kinds of information per pixel:

CHROMINANCE and *LUMINANCE*

CHROMINANCE is defined as the color part of the signal, relating to the hue and saturation.

LUMINANCE is the brightness of the signal — the scale is from black, through grey, to white. A black and white television is simply dealing with the luminance signal, and cannot pick up the chrominance.

There are different ways to encode chrominance and luminance information; the most common is through combinations of the colors **Red, Green and Blue** (known as R, G, and B). Both chrominance and luminance values can be computed from the relative quantities of a signal's R, G, and B.

Like digital audio, a scale had to be created for digital re-creation of the amounts of Red, Green and Blue. And like the amplitudes at each sample for music digitizing, the scientists determined that for each pixel 8-bit samples taken each of Red, Green, and Blue would faithfully re-create a digital version of an analog image for broadcast on television.

> FACTOID #5 • Although a broadcast-quality master is generally accepted in the United States as having 525 horizontal lines by 640 pixels per line, the resolution that is considered "film quality" is much much higher: over 3,000 horizontal lines with more than 1,000 pixels per line. When broadcast video is digitized, 8-bit color for **each** of the red, green, and blue signals is determined sufficient (256 shades per color); film quality is roughly equivalent to 12-bit color (4,096 colors each).
>
> Although film projects at 24fps, rather than the 30fps for video, a digital FILM image would take up 100 times as much memory for the same amount of time. This is one reason why until recently, computers were not powerful enough to perform computer graphics that are usable in theatrical films. This also accounts for the amazing clarity of High Definition Video — it is a special broadcast video signal format with over 1,000 horizontal lines, and 1000 pixels per line.

Let's pause for a moment to do some quick math.

For DIGITAL AUDIO on compact discs, we have already shown that the computer is sampling the music 44.1 thousand times a second, and at every sample, a 16-bit "chunk" is memorized. That means for every second of music:

$$44,100 \times 16 \text{ bits} = 705,600 \text{ bits per second.}$$

Remember that 16 bits is equal to 2 bytes, so let's do this again:

$44,100 \times 2 \text{ bytes} = 88,200 \text{ bytes per second} = 86.1 \text{ Kbytes/second}$ — per channel (remember: stereo audio has TWO channels of sound).

> FACTOID #6 • If you examine the mathematics here carefully, you may have noticed that the calculations appear wrong. By what rule of metrics is 88,200 bytes equal to 86.1K bytes?
>
> Regular metric rules would make 88,200 bytes equal to 88.2K, but the computer industry has slightly modified the rules to work better with binary mathematics. Although a thousand Hertz is a kiloHertz (KHz), and a thousand grams is a kilogram, a thousand bytes *is not* a kilobyte. In fact, **1024 bytes is a kilobyte.**

To play a full minute of stereo digital audio on a compact disc...

86.1K x 2 x 60 seconds = 10,332 Kbytes/minute = 10.1 Megabytes/minute

. . . you need over 10 MB for every minute . . .which means that your average Compact Disc is simply a memory disc holding 60 minutes worth of digital audio, which equals about 600 megabytes. Because it cannot be erased, or used for anything else, we often forget that a Compact Disc is a piece of storage, not that much different from floppy disks or hard disks.

But if you think digital *audio* takes up space, take a quick look at digital *video*. For each of our 307,200 pixels, we need 8 bits of information for each of the red, green and blue colors; that's 24 bits (or 3 bytes) *PER PIXEL*. (Actually, many video systems use 16 bits — 8 bits for luminance and 8 bits for chrominance, but let's ignore that for now.)

307,200 pixels x 3 bytes = 900 Kbytes PER VIDEO FRAME*.

Now, you need that information moving at 30 frames per second. . .

900K x 30 = 26.4MB/second = 1.54GB/minute

. . . which is more than 150 times as large as digital audio.

Because digital video is so HUGE, there are only a few ways to make it fit in a cost-effective storage medium. Among the most common methods:

1) *Make the picture smaller* — if you must retain the full broadcast quality, you might want to shrink the image down — an image 1/4 of a screen takes 1/4 of the memory.

2) *Decrease the frame rate* — moving video at 30 frames per second takes up three times the memory as moving video at 10 frames per second. Although it is generally an unattractive alternative for editing, displaying fewer frames per second is a viable way to save space.

3) *Decrease the color quality* — rather than store the 8-bits for each R,G, and B, decrease the smoothness of the color palette. For NTSC video, every "bit" you save off the R, G, and B signals increases your storage by 170K/minute. Making the image black and white (removing all chrominance) saves even more.

4) Balance the number of bits allocated to each of the R, G and B colors to maximize efficiency of chrominance and luminance ranges. This is kind of technical, but there are "tricks" that can be done to get better color out of the same number of bits. For example, if you have 12 bits of color, instead of giving red 4 bits, green 4 bits and blue 4 bits, you might give red 4 bits, green 6 bits and blue 2 bits. For mathematical and perceptual reasons, this yields better ranges of color.

5) Use some form of VIDEO COMPRESSION. Video compression is a way to take the digitized information about an image and encode it in such a way as to take up less space. How good a particular technique of compression is often depends on the KIND of images that are being compressed. Over the past few years, a number of compression techniques have been pioneered, in particular, JPEG, DVI, MPEG and Wavelet.

All the companies manufacturing nonlinear editing systems that utilize digitized source material must deal with all of these parameters in designing their system. The remaining factors — image size, resolution, color range, compression — must be dealt with in terms of the limitations of the system's memory storage, and computer processing power. Ultimately, the question always must be answered, "How good does an image have to be for offline editing?" The answers seem to vary depending on who you ask — the range being from just about "D1 quality" to something better than "recognizable."

The earliest versions of the digital systems began with 15fps playback of small images, digitized at 4 bits. After only a few years of development, all frame rates extend to 30fps, sometimes with 2-field recording, and compression techniques are much better. Already demonstrated are better-than-VHS video quality digital images for offline, and full (pardon the expression) "broadcast quality" for online.

* Remember: 921,600 bytes is *not* 921.6 KB because a kilobyte is actually 1024 bytes, not 1000. For more information on this, see Appendix A - Metric Numbers. **Also very important:** "MB" means megabytes, however this is different from "Mb" which means megaBITS — obviously a drastic difference in the amount of data represented.

COMPRESSION

Compression is simply the task of taking something BIG and squeezing it down into something small. Today, when you say "compression" people often assume you are talking about *digital video* compression — however true that may be, there are many other things that need to be compressed: analog video, digital computer files, digital video . . . .

Why would you want to compress something? If you want it smaller, of course. You might want it smaller because it takes up too much physical space or because it is too big to easily move around. Since digital storage isn't cheap, and is generally priced per byte, if you can fit more files in the same space, you are saving money. There are many computer programs that compress and de-compress files: Stuffit Deluxe™ (from Aladdin), Disk Doubler™ (from Salient), to name two.

The other major reason to compress something is that information must pass through a limited pathway. If computer data is like cars on a highway, the data *bus size* is the number of lanes in the highway. When a busy highway closes a few lanes to cars, you end up with a traffic jam. Computers often manipulate digital video but may have trouble moving all that data fast enough down the tiny bus "paths" that connect different parts of the computer. This data must be compressed to move smoothly and quickly.

Computers have limitations too. Hard disks, optical disks, floppy disks — all storage devices have certain maximum speeds that data can be "streamed" off of them. Hard disks are currently the fastest (at around 5MB per second), MO disks are medium (500K per second), floppy disks are slow (about 80K per second). So it doesn't only matter how much a storage device holds, it is very important how fast that device can move the data.

Once the data is moving off the storage device, in still may be limited by the pathways around the computer (like miniature highways) since they have a limited number of "lanes." And the computer's chips can only handle so much work. For many other computer applications, the storage device, the pathways and the chips are all of sufficient capacity. But recall that digital video is HUGE and has special needs.

A full broadcast-quality picture requires about 1MB per frame, which (at 30fps) translates into about 30MB per second.

PROBLEM: Our fastest affordable hard disk can only "play" at about 5MB per second. Through some simple mathematics, it can be seen that you'd need to compress this video picture six times (written as 6:1) to get it to play from a computer's hard disk. And that assumes the rest of the computer can handle it. Most computers can't handle good quality video anyway; often the processor and the pathways prove to be the bottlenecks more than the storage device.

TECHNIQUES

There are many strategies by which an engineer can approach the compression problem. You want to shrink down the data as much as you can while losing as little information as possible. If you don't want to lose ANY information (called *loss-less* compression), you can't compress very much. If you can afford to lose *some* information (called *lossy* compression), you can shrink data down much more.

A CRUDE BUT EFFECTIVE DEMONSTRATION:
Original: The quick brown fox jumps over the lazy editor.
Lossless Compression: Thequickbrownfoxjumpsoverthelazyeditor.
Lossy Compression: thquikbrnfxjmpsovrthlzyedtr

The original sentence is 2.88 inches long. By taking out the spaces and squeezing the typeface down, we have effectively shortened the sentence without losing anything; all the information is there — all the letters, the capital, the period. Now, however, it is only about 1.75 inches long. This is lossless, and the sentence has been compressed by 40%. For the lossy compression, we not only removed spaces and used a squeezed typeface, but obvious letters and punctuation were also lost. Notice however that even with 57% of the original sentence thrown away, it is still fairly easy to figure out what it said.

[For an expanded discussion of compression styles, see Compression Made Easy - Examples, on the following pages.]

Many industries are interested in compression: to play video over telephone lines or the internet, or movies off of compact disks, or to broadcast HDTV all require the video to be compressed and each has specific compression needs. Although there are many compression techniques (*e.g.* Indeo, Cinepak, Px64, H.261, *etc.*), two are of particular interest to digital video (nonlinear) editing systems.

JPEG: (pronounced "Jay-Peg") was created in the late 80s by the Joint Photographic Experts Group. They wanted to compress color images and their method is particularly good at doing that. The computer groups pixels into blocks of 64 (8 by 8), and then makes some generalizations about each area. Because it is only looking at a single frame, this is called an **intra**frame method. Today, JPEG is commonly used to compress video around 25:1

COMPRESSION MADE SIMPLE
EXAMPLES

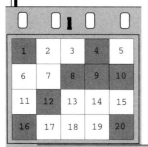

I want you to examine the 20 pixels in these imaginary frames of digital film. Each has an identification number. Each is a color: either white or black. I have 3 frames of film similar to this one — a tiny clip from a movie — and I want you to communicate the patterns of colors in these 3 frames over the phone to me in Santa Cruz, and you're being charged for the duration of your long distance call. Please describe the pattern to me as quickly and precisely as possible. Ready? What might you say?

No Compression

"Hi Mike. **Frame 1**: 1-black, 2-white, 3-white, 4-black, 5-white, 6-white, 7-white, 8-black, 9-black, 10-black, 11-white, 12-black, 13-white, 14-white, 15-white, 16-black, 17-white, 18-white, 19-white, 20-black.

Frame 2: 1-black, 2-white, 3-white, 4-white, 5-white, 6-white, 7-white, 8-black, 9-white, 10-black, 11-white, 12-black, 13-white, 14-white, 15-white, 16-black, 17-white, 18-white, 19-white, 20-white.

Frame 3: 1-black, 2-white, 3-white, 4-white, 5-white, 6-white, 7-black, 8-black, 9-white, 10-black, 11-white, 12-black, 13-black, 14-white, 15-white, 16-black, 17-white, 18-white, 19-white, 20-white."

29.2 seconds.

Lossless INTRAframe Compression — Run Length Encoding (RLE)

"**Frame 1**: 1-black, 2,3-white, 4-black, 5-7-white, 8-10-black,11-white, 12-black, 13-15-white, 16-black, 17-19-white,20-black.

Frame 2: 1-black, 2-7-white, 8-black, 9-white, 10-black, 11-white, 12-black, 13-15-white, 16-black, 17-20-white.

Frame 3: 1-black, 2-6-white,7,8-black, 9-white, 10-black, 11-white, 12,13-black, 14,15-white, 16-black, 17-20-white."

18.5 seconds. [in this case, about 3:2 compression]

Comments: This an efficient way to compress a file and not lose any information. You'll notice that big areas that are all the same color will compress more because you just say something like "one through 500 are all white" and you've transmitted a lot of accurate data. When every pixel is different, the compressed image will not be much smaller than the original.

INTRAframe Compression — "kind of like JPEG"

"Look at vertical pairs; I'll give the top number and then average the pair: **Frame 1**: 1-grey, 2-white, 3-grey, 4-black, 5-grey, 11-grey, 12-grey, 13-white, 14-white, 15-grey. **Frame 2**: 1-grey, 2-white, 3-grey, 4-white, 5-grey, 11-grey, 12-grey, 13-white, 14-white, 15-white. **Frame 3**: 1-grey, 2-grey, 3-grey, 4-white, 5-grey, 11-grey, 12-grey, 13-grey, 14-white, 15-white."

[14.8 seconds. In this case, about 2:1 compression]

Comments: JPEG works by looking at blocks of pixels within the frame and averages their values. This is like having fewer pixels, but when I go to draw my own frame from your JPEG data, I will be close but I'll have to guess to fill in missing info (because I was averaging, when my result is grey, I can't tell which of the pair of blocks was white and which was black; I have no way of recreating the pair accurately). Consequently, this clearly shows a **lossy** method of compression — as is both JPEG and MPEG.

INTERframe Compression — "kind of like MPEG"

"**Frame 1**: 1-black, 2-white, 3-white, 4-black, 5-white, 6-white, 7-white, 8-black, 9-black, 10-black, 11-white, 12-black, 13-white, 14-white, 15-white, 16-black, 17-white, 18-white, 19-white, 20-black. **Frame 2**: same except 4,9, and 20. **Frame 3**: same as frame 2 except 7 and 12."

[10.9 seconds. In this case about 3:1 compression.]

Comments: MPEG works great with moving picture because, except for at edits, each frame is very similar to the frame before it. MPEG encodes a keyframe every 10 or so frames of film, and then simply describes the differences between that key frame and the following frames. This is excellent for playing back movies, but is problematic in editing because wherever you make a cut, that frame may not be a key frame and thus the computer would have to pause and re-calculate what that precise frame would look like. According to compression scientists, this poses a hurdle to MPEG editing systems. (*Note: this particular interframe example is NOT lossy. Real schemes are obviously more complex and involve combinations of techniques.)

with marginal quality loss; even higher compression ratios (from 40:1 to perhaps 100:1) can be used, but with significant sacrifice of image quality. Most nonlinear systems use JPEG.

MPEG: (pronounced "Em-Peg") is a compression method specifically tailored for moving pictures (a kind of spin-off of JPEG, brought to you by the Moving Pictures Expert Group in the early 1990s). MPEG does not simply look around the single frame and compress it. It also looks at adjacent frames to see which pixels are changing and which are mostly the same. This is called **inter**frame encoding (and, if you were wondering, uses "bi-directional prediction motion interpolation"). For example, if your face is on the screen, and you're just talking, probably the only part of the screen that is changing significantly is your mouth — the rest of you remains pretty much the same. By concentrating on the interframe changes, MPEG greatly reduces digital video data — at least 3 time better than JPEG, and as MPEG variations improve, perhaps as much as 10 times better than JPEG.

Another feature specific to MPEG is that it is also designed to handle sync audio. The JPEG techniques only deal with pictures.

JPEG and MPEG are not products — only standardized techniques. Before these were standardized, companies developed their own ways to compress video. With standardization, editing system manufacturers and computer chip manufacturers utilize these methods in their designs. For example, when manufacturers apply the JPEG scheme they generally all refer to their unique proprietary products as using *Motion JPEG.*

The JPEG scheme for compression is commonly found in chips made by C-Cube Microsystems, as well as from companies like LSI. The Intel company has developed its own compression techniques for its Digital Video Interactive (DVI) chips. DVI compression is more common to IBM and IBM-clone computers. This compression technique is proprietary, but it can incorporate the ability to encode/decode in JPEG and MPEG, as well as via other methods.

JPEG is considered the compression scheme of choice for applications that are concerned with still images: desktop publishing, electronic photo processing, digital scanners, color laser printers, *etc.* It is also an inexpensive and simple method — the same JPEG chips that compress video can be used in reverse to decompress. This is called a *symmetrical* technique. The "real time video" mode of DVI is another symmetrical technique.

MPEG is an *asymmetrical* technique. This means that compression and decompression are handled differently, by different sets of chips (or software). This makes MPEG well-suited to different kinds of applications:

playback of pre-compressed video from digital videodiscs or CD-ROM, *etc.* MPEG is designed for moving images and sync sound, but has some trouble handling edits (an edit to a new shot introduces some pretty radical interframe changes). Although MPEG can compress better, it is still somewhat expensive and not yet necessarily viable for editing systems.

If you are not a computer genius, it is difficult to sort out the technical pros and cons of each of these methods. And for the most part, you shouldn't have to. JPEG was first, and is thus a little bit more established and refined than MPEG. MPEG is newer and thus solves many of the inadequacies of its predecessor. So although JPEG is currently considered a totally appropriate editing style, MPEG and DVI compression, once ripe, will be capable of even better performance. Unlike the *discrete cosine transform* (DCT: the mathematical basis for JPEG, MPEG and DVI) newer trends in compression involve higher mathematics like *fractals* and *wavelets*. AWARE Inc., produces chips that use a *discrete wavelet transform* (DWT) — a method that looks at an entire field rather than just blocks of pixels — to get much higher compression rates with even better image quality. At present ,wavelet methods are utilized exclusively in the ImMIX editing products.

There is a great deal of discussion about the variations on the MPEG standard, in particular, MPEG-1 and MPEG-2. MPEG-1 was designed to handle compression of material being played at 24fps (like film), 25fps (like PAL video) and 29.97fps (like NTSC video); it was standardized by the ISO in 1991. What was not considered was material moving at variable speeds, or in reverse. While MPEG-1 is appropriate for many playback situations, it is clearly not ideal. Consequently, MPEG-2 was designed to solve many of these problems and to handle higher output resolutions. However, both formats have distinct advantages and it should not be assumed that by definition MPEG-2 *replaces* MPEG-1. (In fact, MPEG-1 actually provides better quality images at certain compression rates.) MPEG-1 is subset of MPEG-2, which allows for MPEG-1 video to be played from MPEG-2 systems.

OPTICAL DISKS

Optical Disks are removable and re-writable digital storage media that encode data magnetically, like hard disks, but are only rendered magnetically active via a laser. Unlike memory (RAM), optical disks are not volatile — they cannot accidentally be erased by removing power; unlike hard disks and floppys, optical disks are not susceptible to accidental erasure or damage by magnetic fields (generated in X-ray machines, from video displays, from the coils in speakers, from motors, and so on).

There are two kinds of optical disks: *magneto*-optical disks, and *phase-change* optical disks. Magneto-Optical Disks (or MO disks) are recorded as follows: first a laser reads the disk to determine the orientation (or phase) of the bits and kind of *melts* a tiny portion of the disk material; then a magnet writes to the disk by changing the phase of a bit in the melted area; and after the spot cools and hardens, a laser verifies the change.

For phase-change optical disks, the disk medium is in either an amorphous or crystalline state, and is recorded and read by a single laser. The key to both kinds of optical disks is that a laser heats up the disk substrate to a point where the molecules are excited and able to have their orientation changed. Because of the relatively long time it takes for the laser to heat the disk, optical disks have much slower read/write times than hard disks. The phase change opticals, made principally by Panasonic, are cheaper and smaller, but cannot be made to store as much as the MO Disks.

MO disks are made by companies such as Ricoh and Sony; the Maxoptics "Tahiti" drive is popular for editing systems. Unfortunately, even the best MO drives are often too slow for video editing; material often must be first transferred to hard disk before editing can occur.

RAID TECHNOLOGY

Because of the access time and data transfer rate limitations of MO Disks and hard disks, other strategies have been developed to increase the functionality of the medium. Called RAID — for Redundant Array of Inexpensive Disks — a number of disks are used together in parallel to simulate a much faster single disk. Note that a RAID can be designed from any disks that are considered "inexpensive" and today most RAID drives are comprised of non-removable hard disk sets.

Digitized material is "striped" across 2 or more disks (4 or 5 are common): little consecutive pieces of data are recorded onto different disks. For example: where an average MO disk might have an access time of 11ms, and a data transfer rate of about 1MB per second, a RAID of 5 MO disks might produce an effective access time of 2.25ms, and a data-transfer rate as high as 10MB per second. Similar degrees of improvement are seen with fixed hard disk RAIDs. High performance hard disk arrays require extremely high bandwidth pathways to move the data — the usual SCSI interface is insufficient for many products. SCSI-2 interfaces that boast about 20MB/second bandwidth are even being replaced with UltraSCSI (sometimes called Fast-20 or double speed SCSI) interfaces that provide 40MB/second maximum data transfer rates. Remember that for many high quality video applications, uncompressed images are 1K x 1K with 8-bits per pixel (about 1MB/frame) running at 30fps (thus requiring bandwidth around 30MB/second). High performance RAIDs with UltraSCSI interfaces are required to manipulate uncompressed NTSC (or D-1) resolution video in real time.

RAID technology has made the biggest impact on systems where playback of high (or broadcast) quality video is required — online systems, and dedicated animation and graphics systems. When hard disk performance does not improve as quickly as market demands, RAID technology can fill it to augment the systems and deliver the data. When bandwidth is limited by other gating technology, RAIDs are not effective and some kind of data compression will be required to radically decrease the size of the video. For the most part, offline nonlinear video systems do not utilize hard disk RAID technology, and only occasionally MO RAIDs.

DIGITAL MEDIA
CHARACTERISTICS

All digital media have certain characteristics that define their usability and performance. These fundamental traits are:

STORAGE CAPACITY
DATA-TRANSFER RATE (maximum and sustained)
ACCESS TIME

Storage Capacity simply means how much data — expressed in terms of bytes, kilobytes (KB), megabytes (MB), or gigabytes (GB) — can be stored on the media.

It is important to note that many digital storage media — in particular magnetic disk media — must be prepared before they can hold data (it is not entirely unlike "blacking" a videotape prior to use). This preparation is called "formatting," and the formatting not only prepares the disk, but actually records some information onto the blank disk. The actual amount of format information depends on how big the disk is. For example, 2.4GB hard disks, once formatted, actually can only hold about 2.0GB; 3.7GB hard disks format down to about 3.0GB.

Some types of media, in particular MO disks, are double-sided. This means that although the disk may hold 1GB, only 500MB is accessible at one time. If two pieces of data must be retrieved around the same time, they need to be on the same side. This dramatically affects the use and strategy involving MO disks.

Data Transfer Rate is how quickly data can be moved onto or off of the medium. It is actually the resultant of a number of processes: the *internal data transfer rate* (how fast the read/write head can get data on and off a disk), the *sustained data transfer rate* (the speed of material through the drive controller). If data on a disk is like water in a bucket, and the bucket has a hole in the bottom, the data transfer rate relates to the size of the hole. With a tiny hole, you can only empty the bucket so fast; a larger hole pours out water much faster. Data transfer rate is important because for video to play, you need to move a great deal of data very fast off the digital storage device. If you can't move it very fast, you need to make the digital memory size of each frame smaller so that you can pump them out at 30fps. There are, of course, various ways to accomplish this (decrease the frame rate, make each frame image smaller, *etc.*). Each of these compromise picture quality to varying degrees.

Some devices advertise very high data transfer rates, but these are generally the *internal* speeds; for smooth video playback you need a *sustained* (or average) rate that is pretty high. This sustained rate is the important item for a storage device. It is also important to note that transfer rate is slightly different for reading (pouring out) and writing (recording in).

Related to data transfer rates are the rates at which the data flows around the computer, through cables and between components. Here you have protocols like SCSI-1, SCSI-2, Ultra-SCSI (or Fast 20) and so on, for cables and connectors; and ISA, EISA, PCI, NuBus for the internal computer pathways. All of these can further limit the speed at which the data flows around the computer.

Access Time is how long a certain device takes to locate a chosen piece of information. Access time is actually composed of "seek time" or the time it takes the disk to cue from one track to another (and with the worst-case scenario a seek from the inside to outside radius of the disk) and "rotational latency" which is kind of how fast the disk spins. On videotapes or laserdiscs, access time is a critical factor, as it determines how "random access" a medium can be. If the access time is slow — like for videotapes — a system needs multiple copies to find a given frame quickly. For laserdiscs it takes only about 1.5 seconds (about 45 frames of video time) to cue to a desired frame.

Computers also need to access material. When you ask for a shot or a file, the time it takes to find it, load it into memory, and present it to you, is generally considered the "access time." This time, however, is miniscule when compared to the shuttling of videotape. For reference, consider that a complete video frame takes 1/30 of a second, or 33 milliseconds. Hard disks can access in less than 10ms.

The size of an uncompressed broadcast-quality video frame is about 900KB. If you need 30 of these every second, then our video-computer must *move* data at a rate of 27MB per second. A quick glance at the specifications of a good hard disk, regardless of how much data it HOLDS (its *capacity*), tells you that it can only move data ("sustained transfer rate") at about 5MB per second. This is one reason why compression techniques are so important to digital video and computers — not simply to make the economics better for video storage, but because most devices simply could not manipulate data that is so large and must be moved so quickly. A JPEG compression of 10:1 does not degrade images too much, but reduces the required transfer rate to a much more easily managed 3MB per second.

QUALITY-COMPRESSION TABLE

As a handy reference, this table outlines some general parameters of image compression. From these figures, you can investigate the limits of a given hard disk or alternate storage device (the *sustained data transfer rate* would need to equal or exceed the "MB/sec" figure) for example. You can also check the viability of different compression schemes (*e.g.* JPEG vs. MPEG). Clearly, between these well-known "quality benchmarks" are a continuum of values and therefore resolutions.

quality benchmark:	"D1" (CCIR-601)*	"Beta SP" or "1"	"S-VHS" or "Hi-8"	"VHS"
compression rate:	none	3:1 - 5:1	8:1 - 12:1	20:1 - 30:1
frame size (KB):	680	120 - 200	50 - 80	20 - 30
MB/sec required @ 30fps:	20	3.5 - 5.6	1.5 - 2.3	0.6 - 0.9
Storage per GB (min:secs):	0:51	3 - 4:30	7:30 - 11:20	19 - 28

Due to the popularity of Avid products and their widespread use, the following table is included here for comparison. It shows Avid's Video Resolution (AVR) compression values along with the minutes of video per gigabyte. The first row shows the storage values for 30fps material and **no** audio. The second row shows the addition of 2 channels of 44.1KHz audio. Note that the addition of audio data reduces the amount of overall storage time available, with the greatest effect at low AVR values, and a lesser effect at high AVR values. The third row gives estimations of image frame size at each value, allowing rough comparisons to the above table and other video products. These compression values allow for image resolutions ranging from about Beta SP quality (on the left) to below VHS quality (on the right):

	AVR 27	AVR 26	AVR 25	AVR5	AVR 4	AVR 3	AVR 2	AVR 1
pix-only	4 - 9	6 - 9	9 - 19	14 - 22	24 - 48	30 - 56	44 - 85	63 - 107
pix+snd	4 - 8	5 - 8	8 - 16	12 - 18	19 - 32	23 - 35	30 - 45	38 - 51
KB/frame	65 - 146	65 - 97	31 - 65	26 - 42	12 - 24	10 - 19	7 - 13	5 - 9

*Note: CCIR-601 is a broadcast-quality digital image with 720x480 pixels at 60 fields per second for NTSC. This is slightly larger than the more common high-quality image of 640x480 pixels, which produce frames of 622KB.

SUMMARY OF DIGITAL STORAGE ALTERNATIVES

There are a number of ways to store digital information, whether for audio, video, or simply data (like text). Here is a quick sampling of popular choices:

Floppy Disk: standard double-density disks hold about 750K; high-density disks hold about 2MB. Transfer rates are slow, around 80KB/second.

Floppy Disks are thin flexible disks covered in magnetic oxide and contained inside a protective sleeve in which it rotates. They are lightweight, cheap, and portable. However, because the protective sleeve is not very strong and needs access holes for the computer heads to get to the disk (through which dirt and dust can also enter), they are not as reliable as other digital media. Their cost and size have made them popular with the personal computer market.

The first floppies were 8-inch disks — in an 8" square package — but those have pretty much been replaced by 5 1/4" and 3 1/2" versions. Because the 3 1/2 inch "diskettes" are in hard casings, and have a little protective "door" over the holes to access the disk, they are more reliable than other versions. Current technology offers standard floppy disks up to 2MB.

Hard Disk: available in sizes from 20MB up to 9GB. Transfer rates are very fast, currently between 5-8MB/second. RAIDs of hard disks multiply these rates many times.

Hard Disks are rigid and protected disks coated in magnetic oxide. They primarily come in three varieties: removable, disk packs, and fixed (like Winchester Disks). Hard disk drives generally contain a number of disk platters (up to around ten), each storing data on both sides. Removable disks are generally more expensive per byte than fixed disks, and can be less reliable because of potential dust-dirt problems and handling shocks. It should be noted that there are different varieties of removable hard disks. There are removable media — these are hard disk cartridges of anywhere from 44MB to 230MB and are made by companies like Syquest and Iomega. There are also portable external hard disk drives. These are small, high volume devices that hold many gigabytes of data and are made by many companies, like Seagate, Micropolis, and Quantum. Then there are the removable drives (R-mags) which are self-contained disk drives that slide into semi-fixed external chasis and hold hard disks presently up to 9GB.

Optical Disk: available in sizes up to 2GB. Transfer rates are slower than hard disks but faster than floppies — between 100KB and 800KB/second.

Optical Disks are a relatively new form of re-writable digital storage. Some optical disks are not erasable (data is literally burned onto them) and are often called WORM disks (Write Once, Read Many). MO Disks are small, about the size of traditional floppy disks, and both portable and safer from damage than magnetic oxide storage. Unfortunately, the slow data transfer rate makes high quality video difficult to move at adequate picture resolutions. There are two kinds of optical disks, differentiated by the technology used to read and write data: magneto-optical (MO) disks and phase change optical (PCO) disks PCOs have more than twice the data transfer rates than do the MOs:

CD-ROM (and Compact Discs): available up to 1.5GB. Transfer rates are standardized at 1.5Mb/second.

The Compact Disc, whether used for high fidelity audio playback (CD), or permanent data storage (CD-ROM), is an easily mass produced, low cost digital media. Compact Discs for music playback were invented by Philips in the early 1980s; later, CD-ROM and CD-I were new formats of digital disk recording for specific applications in the computer world and the interactive educational/enter-tainment markets respectively. CD-ROM can be used to store data — from encyclopedias and dictionaries, to phone books, all easily accessed via a computer. 1.5 GB can be 70 minutes of high quality stereo audio or about 100 volumes of text. In the past few years, the price of CD recorders has dropped substantially, and the cost of blanks is low. This is making them ideal choices for the long-term archival of data.

Digital Audio Tape (DAT):   storage depends on tape length.

DAT was invented in the mid-80s as an alternative to the strictly playback functionalities of Compact Discs. DAT, though linear, is an extremely fast and high quality recordable media, ideal for both professional and consumer audio record-ings.

Digital Video Tape: storage depends on tape length.

Digital video can be recorded linearly on tape, like traditional (analog) video. Signals can be encoded in either component or composite, on small 19mm cassettes. Signals are encoded in an 8-bit format. There are many formats for recording on digital video tape, all of which affect the quantity of available source: for example, D1 tapes are available in 34 or 76 minute lengths; D2 tapes record up to 208 minutes.

SPECIAL EFFECTS
A BASIC INTRODUCTION AND OVERVIEW

Special Effects, or "SFX" or "F/X" are manipulations of images used in editing. They may be simple, like fades or dissolves, or they may be complex, like morphing. An effect may combine shots or manipulate a single shot. Either way, there are a few basics that may help you understand the nature of video and film effects.

SIMPLE EFFECTS

Simple effects are dissolves, fades, superimpositions, or wipes. In short, these are all different ways to replace one picture gradually with another. In the case of a *fade*, one picture is being replaced with a "black" picture (or the other way around, in the case of a "fade up"). A *dissolve* is when an outgoing picture fades out at the same time as an incoming picture fades in. One way to visualize what is happening is to have one picture track overlapping with another picture track, and then you slowly (or quickly) switch from one track to the other. The classical type of dissolve is a "center point dissolve" which means that when the outgoing picture is halfway gone, the incoming picture is halfway in. A *superimposition* (or "super") is when you see both frames at the same time for an extended period of time. A *wipe* is when one picture replaces the next via a moving geometric shape — a circle, a box, a line, whatever. Which ever type of simple effect is desired, the same principle applies: you need to have independent control of two separate channels of video and then some way of combining them to get output.

COMPLEX EFFECTS

More complex effects are *keys*. A key is where a shaped hole (called a "matte") is cut into a background and that hole is filled with another picture. The shaped hole, however, does not always need to be a solid geometric shape. Any way you can choose some portion of an image to separate it from the rest is a viable way to cut out a key matte. A very common way to cut out a matte is to have a computer locate a single color ("chrominance") or brightness level ("luminance") in your picture, and have *that* cut out. If you want to cut me out of the background in a picture, it is easiest if you make the background easy to identify: a evenly lit solid color. It would also help if that color is no where else in the picture. Consequently, when you intend to do these "chroma keys" set up a brightly lit green (or blue) screen behind me, and then cut out the green (or blue). The classic example of this is the local weatherman standing before a map during the morning broadcast. He is actually standing before a green screen and watching himself on a little side monitor to see where he is pointing. Sometimes, if he accidently wears a tie with a little green in it, you can see the map in his tie. Again, you need two video channels to do a key: one for the background (it this case, the map) and one for the foreground (the weatherdude). But you also need the shaped hole (remember, the matte) to put in the middle. In computers, this is called an alpha channel.

Another complex effect is called a DVE, short for *digital video effect*. With these, a frame or series of frames are independently manipulated with respect to the background: the entire frame may be resized (bigger or smaller), moved (left, right, up, down), rotated (spinning) and flipped and so on. It is important to note that these transformations are not just in X-Y coordinate space (the plane of the flat screen) but in X-Y-**Z** (actually *into* the screen).

TITLES AND CHARACTER GENERATION

These are exactly how they sound: the system allows the user to type text onto the screen, adjust font and size (and perhaps other characteristics), and add effects (like drop shadows). While the user may not realize it, these titles are actually keys, being cut out of the sequence background — and consequently can themselves sometimes be filled with moving video.

MOTION EFFECTS

Motion effects are any time you change the speed of the film or video and incorporate that new footage into the sequence. This may include *speed-ups*, *slow-mo*, and *freeze-frames*. There are special ways that editors may want to create motion effects: perhaps you have a shot of me walking to the door, but I seem to get there too quickly. You want to do a slow-mo, but you don't want it to look that way; you simply want the 3 second shot to take 3.2 seconds. It is as if you have a 3 second shot that needs to fit in a 3.2 second hole — this is called *fit-to-fill*. Motion effects are generally easy to create electronically in offline, but difficult to re-create when conforming film and necessitating film opticals.

PAINT

Computers are readily able to paint the pixels you see on the screen. This might be drawing an entire work of art from scratch — starting with a blank screen and filling it with your painting. If you are an animator, you would then have to paint every single frame (remember: film is 24 frames per second); a daunting task. Alternately, you could use your paint tools to paint an existing image, perhaps a film or video image. In this case, you are only changing colors and modifying the frames. This is used to produce many effects in film and television, among them *rotoscoping* (frame by frame additions over live action): this is how the phasers in *Star Trek* are added. Painting software is often referred to as 2D paint because you are only modifying the flat image of the screen. When software allows you not only to paint frames but to control the frames and the objects within them over time, the paint system is said to also do *animation*.

CGI

Rather than paint existing live action objects (or modify them), sometimes it is cheaper or necessary actually to draw a new object in a kind of architectural design program. CGI or *computer generated images*, are the complex 3D objects that designers create When designed, these CGI objects are generally only the skeleton of an object (called a "wire frame"). Once this is completed, a skin needs to be applied to the "bones". The skin can be painted on (via a paint program) or *texture mapped* (when a painted object is literally wrapped around the wireframe). Then the realistic 3D CGI object can be placed into a number of different environments and animated. This way, rather than draw each frame of an animation (the old 2D way), the object need only be moved around the frame over time and each frame is like a snap shot of that activity. Many television commercials use CGI today to animate Coke bottles, beer cans, bears, cars, and so on.

PLUG INs

A *plug in* is a software module that has been written in a standardized format, and that can be used inside a more complex software product. For example, Adobe Photoshop pioneered the use of plug ins: they packaged their image manipulation product with some special effects (blur, sharpen, *etc.*), and then they allowed third-party vendors to create their own (like "lens flare" and "mosaic") and *plug* them into Photoshop. Editing software adopted this idea and allowed for these Photoshop plug-ins, as well as many others, to fit into their special effects functionality. When a product works with plug-ins, the user has a great deal of flexibility to expand the type and range of effects.

EFFECTS AND NONLINEAR EDITING

Because not all nonlinear systems can handle more than one video "stream" going through the pipe at once, effects may have to be "rendered". This means that each frame of the effect must be calculated and generated out of real time. The effect is literally built by the computer frame-by-frame and stored on a hard drive. While it may be transparent to the user, the system cuts in this new rendered effect in the place of the original shots. (This is quite parallel — albeit automatic — to ordering film optical effects, and then cutting them into the workprint or negative.) Rendering is required when the computer does not have the design or power to preview the effect live (see Chapter 6: vertical nonlinearity). Since the editing system is storing these rendered effects along with the other source material, creating many effects while editing sometimes needs to be a storage consideration as they can take up considerable hard disk space. Also, since the effect is playing back from a new rendered source, if you change any part of the effect, the entire thing needs to be re-rendered. This will take some time and may discourage editors from performing many effects, at least until the sequence is more completed.

CHAPTER 4

EDITING SYSTEM PRIMITIVES

WHAT IS AN EDITING PRIMITIVE?

The term "editing primitives" derived originally from computer programmers designing editing systems. Before a program (software) can be written, the system's designer needs to identify what the system will do, and in what way it will do it. These are the "primitives." There are only so many things an editing system needs to do. Editing is pretty simple — cutting and pasting. There are variables, and there are a number of computer functions and database manipulations, but most editing systems deal with similar concepts in pretty much the same way.

COMPARING SYSTEMS

Although it seems that every editing system is unique, they all share certain commonalities that once understood need only be "translated" when switching from system to system. Once one editing system is known, it becomes considerably easier to learn others.

Below is a list of common features culled from manufacturers' brochures. In product literature it is not uncommon to find that many of these functions are listed as if they were unique to that particular product (like "nonlinear", "frame accurate", *etc.*), whereas in reality most ALL systems are:

❏ Flexible — can be used on projects from commercials to feature films.
❏ Fast — at least faster than cutting on either film or tape.
❏ Simple to learn — with simple "intuitive" interfaces "designed" for film editors
❏ Cost-effective (if used correctly).
❏ Reliable — after a few years in the field, they work out seriously debilitating software bugs, and very rarely screw-up.
❏ Designed with at least 2 separate channels of audio.
❏ Able to trim or extend an edit by one or more frames.
❏ Able to undo and redo at least *some* edits/functions.
❏ Able to handle drop- and non-drop timecodes.
❏ Able to do film- and video-style editing.
❏ Able to copy a cut sequence (like creating dupe-workprint).
❏ Able to output lists for both 24fps and 30fps projects.
❏ Able to control a videotape deck for offline videotape assembly of the master.

This list is not a special set of features but rather the *defining* characteristics of professional nonlinear editing systems. Just by introducing a product that is an "electronic film" or "nonlinear video" editor, you pretty much have to satisfy the above-mentioned criteria.

But there *ARE* factors that set them apart. Systems are different: they look somewhat different, they have different computers using different source media and interface styles, different functions and abilities, with strengths and weaknesses — usually related, at least in part, to COST. The following checklist is a partial summary of the questions that prospective users and buyers should have answered when determining the proper nonlinear system:

INPUT/OUTPUT

❏ How much source material can be loaded and accessed at one time?
    What is the picture quality of that material?
    Can you mix levels of resolution within a project? Within a sequence?
    How many degrees of resolution are there?
❏ What options are available for alternative film sizes and speeds (IMAX, 30fps film, 3-perf, 16mm, 8mm, 60fps, *etc.*)?
❏ How much time does it take to "log" and/or input the source material?
    Can logs from telecine be input automatically?
    What format of telecine logs do you take?
    Does source material even need to be logged?
    If not, how do you access dailies? and at what point in the process can the logging be done?
    Can you log remotely? What are the options for doing so?
❏ If a mistake is made in the log-database (timecode-key number relationship), can it be corrected after material has been edited — without requiring re-editing?
❏ Can the system input an EDL from another editing system?
❏ What options are available for alternative source and record machines (1", S-VHS, Beta SP, *etc.*)?
❏ How long does it take to assemble a videotape of the master cut? How "frame accurate" is this assembly? Does the system output a videotape using the digital source material only, or can it do a tape-to-tape assembly? A-mode? B-mode? C-mode?
❏ If the system is editing via timecode with plans to convert to film edge numbers, what "rounding" corrections does the system make? Is it counting/editing at 30fps or 24fps? Are the implications of this acceptable?

❑ What formats of EDL (Edit Decision Lists) are available (CMX, GVG, Sony, *etc.*) and on what size floppy disks?

   What information is included in the EDL? *e.g.* GPI? Notes?

❑ What kinds of film lists can be output? Pull lists? Change lists? Optical count sheets?

❑ Does the system account for re-using film material? Does it account for negative cut margins? How are these flagged and coordinated with potential requirements for reprints and opticals?

❑ How much source or edited material MIGHT be lost if the system's power is suddenly shut off?

   Is source or edited material backed up and protected?

   How is this accomplished? Is this easy to handle? Time consuming?

   How easy is it to restore?

❑ What file formats are supported for importing and exporting?

EDITING FUNCTIONALITY

❑ How is most user input accomplished on the systems — keyboard? mouse? trackball? touchscreen? touchpad? *etc.*

   Is the physical interface very complicated? comfortable? efficient?

❑ Are there more than 2 channels of audio?

   How many can you edit with at one time?

   How many can you input and output at once?

   If fewer can be output than edited, can the multiple tracks be mixed down for output?

   How many channels of audio can you monitor at once?

❑ Is the system capable of EQ?

❑ Can audio be easily moved from one channel to another? Mono or stereo?

❑ Is audio level remembered by the system? How are they adjusted? "Rubber bands"? Console?

❑ What are the effects (wipes, fades, dissolves) capabilities?

   Are they implicit in the system?

   Do they require additional peripheral equipment? or cost?

   Do they have to be rendered? How long will they take?

   Does the system accept plug-in effects from other software products?

   What about motion effects (like speed-up, slo-mo, fit-n-fill and freeze frames)?

   How are these represented in the EDL and film lists?

❏ Can you play the *virtual* master in non-play speeds — variable fast, slow, forward, reverse, frame-by-frame? How does this affect audio pitch and can that be controlled?

    Can you shuttle source and master material with a knob?

    How smoothly can you vary the speed in the playing of material?

❏ Can you examine the outgoing and incoming frames at an edit point at the same time?

    Can they be adjusted together and if so, can they be interlocked and scrolled at play speeds?

❏ How easily can the system perform a spit-edit (also called an *overlap* or *L-cut*)? This is a very common function that should be effortless.

❏ Is there any limit to number of events (edits) in a cut sequence?

    Is there any limit to the number of sequences that can be cut?

    Is there any limit to the number of log entries that can be used?

❏ How difficult is it to add additional material, or late arriving material, into a cut sequence?

    Can a shot from one project be easily moved into another project? What is the process?

❏ Can the digital source material be easily switched (do you have to close up and turn off the system to swap source drives or media)?

❏ Can edited "master" material be moved to "source?"

SYSTEM/CORPORATE

❏ Is the system based on proprietary hardware, or open platform?

    What happens when the hardware is updated?

    Is there an upgrade path to other products, future updates, or future products?

❏ Is technical support available? Does it cost extra? what are the hours?

❏ How easy is it going to be to find a trained base of editors for this product? How many systems are currently installed?

❏ Are software updates free? If not, how much do they cost?

❏ How easy or difficult is the system to set up and cable?

    Is the physical configuration otherwise limited? Cable lengths, *etc.*?

❏ Is it available as a bundled hardware plus software product that can be installed on a customer-supplied platform (SGI? Mac? PC?)

Recognize that "more" isn't always better. A system that does MORE is usually more complicated, although not necessarily so. What you want is a system that is correct for the project being edited.

From a facility standpoint, a more powerful system means more flexibility in the KINDS of projects that can use the equipment. From an editor's standpoint, there must be functional ease as well as compatibility between the system and the particular production.

EDITING PREPARATION
HOW ELECTRONIC EDITING FITS INTO THE POST-PRODUCTION PROCESS

One significant difference between electronic post-production and traditional linear videotape editing is the time involved in preparing for the actual edit session. Because of the databases implicit in many electronic nonlinear systems, and because many systems do not edit on a coded media, some work usually must be done after the shoot (and after telecine, for film), and before the cutting can begin.

In linear videotape editing, grab a 3/4" tape and you can edit anywhere, immediately. Pop your tape in a source deck, and start watching and editing.

On film, there is already a common infrastructure set up in the preparation to edit. Lab reports, code numbers, key numbers, scene/take breakdown … the move to electronic nonlinear does not radically add to the preparation time to edit. Of course, it can be frustrating to have the dailies on film, and still have to wait for telecine, digitizing or dubbing to 1/2" videotape or videodisc, and then logging the tapes or discs. But in many ways, preparation and organization have always been part of the job.

Linear videotape editors can be stunned by the sheer quantity of steps a film editor has to go through to begin editing.

The major difference between editing electronically and editing on film is that dailies do not arrive on workprint, but rather on some sort of video media — usually 3/4" tape. However, unlike linear videotape bays, there is no 3/4" deck set up for editing. With most systems the 3/4" deck is the vehicle by which the dailies are transported into the edit system. Through this tape, source loads are digitized onto hard disks or optical disks.

The golden rule for nonlinear editing (or more correctly, for real-time previews) is that all the source required to play back a given length of a sequence must be loaded and accessible to the system at one time. Because hard disks are so large today, and because you can effectively gang several disks together (in towers, in SCSI chains, *etc.*), many productions can fit all their source material onto a given system at once. Thus there would never be any need to break the source into loads for editing. However, if the system's source capacity is less than the total volume of source required, material must be broken down into editing loads from the original 3/4" or Beta SP source tapes.

Not all nonlinear editing systems require so much preparation. Any system that does not require a database will be particularly simple to approach and begin using. A system that edits on a media where codes (either code numbers or timecodes) can be read DIRECTLY from the media via the machine controller (like videotape editing, for example) will also require minimal if any preparation. The great desire for many electronic systems is to be more like videotape editing than film in this regard — to be completely "stand alone." They would want you to be able to walk into an edit bay with a tape, and some time later, walk out; with minimal if any special logging or preparation.

Still, it is generally acknowledged that the preparation you do before editing makes the edit and the completion of the project much faster. The time-consuming parts of post-production have merely been shifted around, so they are not readily comparable to the former ways of working.

For many systems, the delivery of dailies on 3/4" tape means another 30 to 60 minute commitment for input and preparation. Sometimes more, sometimes less. Following this, the edit begins and moves much faster and smoother than with any film or video session. The time savings here are often so dramatic it is embarrassing. It can appear that the edit was effortless. Certain re-edits that used to be difficult are now simple. On the other hand, tiny changes that used to be minor might now take longer. Once the edit is done, for example, there is still the process of recording those decisions onto a piece of videotape for presentation. In both film editing and videotape offline, finishing the edit can be immediately followed by screening or distribution. Now a new time-consuming element is added to the process. Tape output.

Is it all worth it? Yes: if the time you save editing is greater than the time you lose logging beforehand and making assemblies after. Or if you can now divide up the labor so that less expensive personnel can log and make assemblies, and the high-priced talent can concentrate on a faster, more efficient edit session. It should be clear that there is no simple comparison of time and money savings when investigating electronic nonlinear alternatives, especially when comparing them to either film or traditional videotape editing.

E D I T I N G   F L O W   C H A R T
T HE E LECTRONIC P OST-P RODUCTION P ATH

The familiar method of working on film is known by most people in the film industry — it is the fact that this knowledge has been so well established for decades that makes changing methods so frightening.

However, electronic tools in the post-production of film can be inserted into the process in a number of different ways, and some of those barely change the flow at all. Still, it *is* possible to entirely revamp traditional means of posting film, and utilize all-new electronic and digital equipment. The degree of change is up to each production.

Following this are two flowcharts: the first shows steps necessary to prepare for an electronic editing session; the second depicts finishing a project after edit decisions are complete and picture is locked. The charts are generalized to present options for editing on any number of digital systems. Remember, this is only *one of many variations* that can be used; television, theatrical films, and commercials will all modify this process based on the desired end-product, budgetary issues and creative preferences.

S P E C I A L   N O T E S

❊ For a **video-only finish**, the second flowchart would end with the online of picture and sound, and the traditional audio sweetening session.

❊ For a **film project using workprint**, prior to telecine, work picture and track would be struck, synced, and coded, as in traditional film work. These, rather than negative, would be transferred in the telecine session. Following the locking of the picture on the digital system (or as each reel is completed), the workprint and track would be conformed to the output film lists. From here, post-production would also follow traditional film methods.

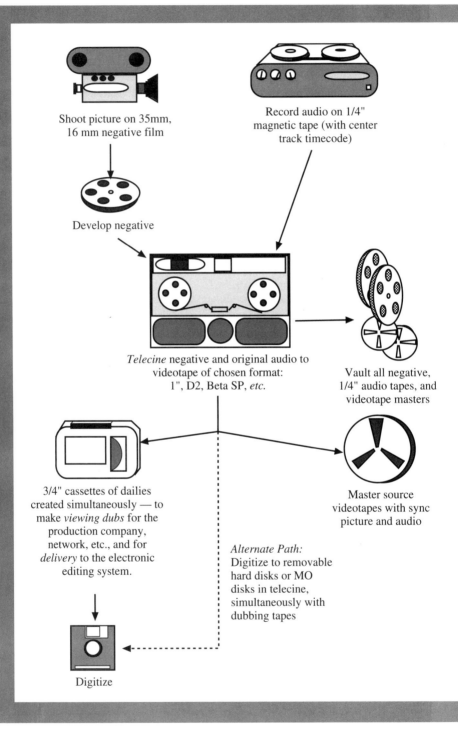

Shoot picture on 35mm, 16 mm negative film

Record audio on 1/4" magnetic tape (with center track timecode)

Develop negative

Telecine negative and original audio to videotape of chosen format: 1", D2, Beta SP, *etc.*

Vault all negative, 1/4" audio tapes, and videotape masters

3/4" cassettes of dailies created simultaneously — to make *viewing dubs* for the production company, network, etc., and for *delivery* to the electronic editing system.

Master source videotapes with sync picture and audio

Alternate Path: Digitize to removable hard disks or MO disks in telecine, simultaneously with dubbing tapes

Digitize

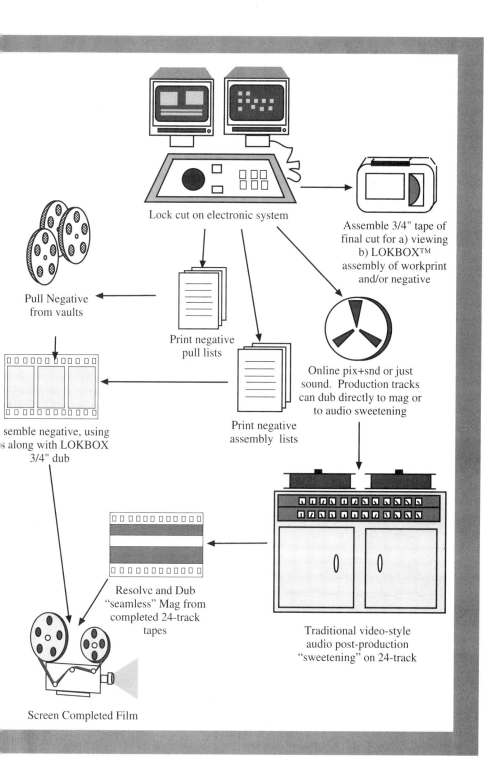

Lock cut on electronic system

Assemble 3/4" tape of
final cut for a) viewing
b) LOKBOX™
assembly of workprint
and/or negative

Pull Negative
from vaults

Print negative
pull lists

Online pix+snd or just
sound. Production tracks
can dub directly to mag or
to audio sweetening

semble negative, using
s along with LOKBOX
3/4" dub

Print negative
assembly lists

Resolvc and Dub
"seamless" Mag from
completed 24-track
tapes

Traditional video-style
audio post-production
"sweetening" on 24-track

Screen Completed Film

C O M P U T E R S

No matter how you look at them, no matter how film-style or graphical editing systems might be, they are all just computers. Consequently, to really have a basic idea of how a system works (or why it doesn't work), a fundamental knowledge of computers is helpful.

A computer is characterized by a monitor (to watch), a keyboard (to type on), a pointing device (like a mouse or trackball), and some kind of box containing a hard disk and fans. Also in this box are computer boards holding many chips, primarily RAM (random access memory), ROM (read-only memory), and a CPU (central processing unit). The physical tangible part of a computer, in particular the circuitry, is called "hardware." The intangible instructions for how a computer works and does tasks is called "software." The manufactured box of hardware and software is called a *platform*.

CPUs

Virtually all computers have at least one CPU, or *central processing unit*. This chip is the brains of the computer — it gives a system its language, and this in turn allows software to have a given style and look.

For personal computers, there are two principal brands of CPUs: Motorola's 68000 series of chips and Intel's 8086 series. Every time Motorola or Intel improves on their technology they update the chip number. A few years after Motorola introduced the 68000 came the 68010, then the 68020, the 68030, and then the 68040. As a general rule of thumb, it is said that computer power and memory double every 2 years. For Intel's CPUs, the number ahead of the "86" has incremented too; Intel had created '286, '386 and '486 chips. Rather than introduce a '586 chip, Intel called it "Pentium". In 1994 when one might have expected a 68050 chip from Motorola, a consortium of companies including Motorola got together and made a special CPU that would be very powerful and work with both Macintosh and PC systems; it was called the PowerPC chip.

For a number of technical reasons, the 68000 series has become common in many newer computer brands and in computers using graphics. Many workstations made by SUN Microsystems and most Apple Macintosh computers (up to their Power Mac line) use 68000-series chips.

IBM and IBM-compatible computers running DOS or Windows are generally '86-based. Their interfaces tend to be slightly less graphical, but they are also considerably less expensive than Macs.

All CPUs have a "clock speed" which relates directly to how fast the chip can perform its calculations. Clock speeds found in 1995 personal computers range from 16 to 120MHz ("MHz" here means millions of clock pulses per second). To do high-end computer graphics, which require an extremely large number of calculations, special CPUs with very fast clocks are required, exceeding even those of PowerPC and Pentium CPUs.

Many editing systems are based on common chipsets and common hardware platforms however that does not necessarily translate into product compatibility.

PLATFORMS

A platform is a product or product family from a computer manufacturer which includes special combinations of hardware (RAM, ROM, CPU) with a specific set of software instructions that determine how the computer "thinks". This software is called the *operating system*. It is the operating system and the hardware that makes the different computer products unique. When software companies have a good idea for a product that they want to design, they must build that product specifically for a certain platform with a certain operating system. Then, when that product is modified from its original design to work on a different platform (with a different operating system) it is called *porting* the software. This task is often quite difficult.

The **Macintosh platform** is now comprised of the "regular" Mac (with the Motorola 68040 CPU) and the Power Mac (with the PowerPC CPU). The Mac operating system is simply called *Mac-OS*. IBM (and IBM-compatible) is a platform type, but a better term today may be "Microsoft-compatible"; we'll call them **PC**s. For PCs, operating systems include DOS (Microsoft's original PC operating system), *Windows*, and *Windows NT*. Windows is a DOS-type operating system that runs on Intel CPUs and is designed for consumers. Windows NT is a totally different version of Windows that runs on high-end CPUs like MIPS, Alpha, PowerPC and so on; Windows NT requires vast amounts of RAM and hard disk space, and consequently is primarily for professionals. *Windows 95* is a consumer-version of Windows NT. **Silicon Graphics** (SGI) makes a high-end platform (*e.g.* Indy, Indigo2, Onyx, Crimson) that was initially designed for very fast professional graphics capabilities, and consequently runs off of powerful MIPS CPUs and utilizes the *UNIX* (pronounced "you-nicks") operating system.

A computer does not need to be in a box made by Apple or IBM. Some manufacturers have chosen to build their own style of computers. A few manufacturers have opted to build their own "proprietary" computers from the ground up. The ongoing computer revolution seems to be demonstrating that open platforms (products that can run on standard computers) are winning out over "closed" proprietary systems — in editing, as well as in many other industries. Open systems *tend* to be less expensive and geared to a wider audience; proprietary systems tend to be more expensive and more finely tuned in terms of features and interface for the function for which it was designed.

MEMORY

A computer's active memory is called RAM, or *random access memory*. This set of memory chips is a workspace where information is quickly accessed and used. Editing itself generally takes place in RAM. It is often called *volatile* storage because it is temporary and will be erased by a loss of power — like if you unplug your computer. RAM needs to be large enough to provide sufficient space to store all information necessary for editing, as well as for the application itself. How big this needs to be is purely a function of the size of editing software and data. A system that can handle a large sequence or production needs a great deal of RAM. This is particularly true for the digital systems that carry actual images as well as computer data. Linear systems move very little data around for editing— they are only machine controllers, shuttling and cueing video decks. While personal computers have typical RAM configurations of between 1.5MB to 16MB, digital video systems require somewhere between 16MB and 32MB.

STORAGE

Computer users make a distinction between forms of digital storage: "memory" is defined to be RAM; "storage" is not chip-based; it is longer term, semi-permanent storage for information. Hard disks and floppy disks are two examples of computer storage. Hard disks are protected from the environment and are considerably safer from power problems than is RAM. Information tends to flow from the hard disk to RAM when you "load up" a project or open a file. When information is saved, it is copied from RAM back to a place on the hard disk.

Hard disks need to hold much more than RAM does. They are not chips the way CPUs and RAM are, but rather magnetic disks that can store high densities of data. Typical internal hard disk sizes range from a couple hundred MB to about a gig; external drives can be purchased in sizes up to about 9GBytes. The largest sizes are most practical for "large" data storage — like digital audio, video and graphics. Most data manipulation and word processing requires considerably less storage. As with RAM, every few years sees a dramatic increase in the density of data that can be stored on magnetic media. Although most hard disks are fixed into computers, newer "removable" or external drives are currently available and are often found in editing applications. Systems are largely designed to use the smaller internal hard disks for editing data and applications; the larger hard disk storage towers, arrays and so forth are saved for the very large video and audio material itself.

While they can be delicate and relatively large when in a removable form, hard disk configurations currently provide for the fastest flow of data to and from a computer. More mobile and smaller, magneto-optical (MO) storage disks can hold more than a gigabyte (a thousand megabytes), but do not data to be transferred on and off as quickly, making them somewhat less useful for the real-time playing of digital video images. Only at lower resolutions can these disks be as functional as hard disks.

Faster than either MO drives or the fastest hard disks is RAM. Information in RAM is moved extremely quickly. Unfortunately, the economics of RAM (cost/byte) are relatively high and the chips are still volatile. All these technologies continue to improve.

ROM, or *read-only memory*, is like a tiny permanent hard disk printed on a computer chip. Like a hard disk, ROM can provide information and instructions to a computer system, but unlike a hard disk, its contents cannot be modified. ROM is fundamental to all computers and thus all editing systems, but has little relevance to the investigation or comparison of systems. More flexible forms of ROM, in particular PROM (**P**rogrammable **ROM**) or E-PROM (**E**rasable-**P**rogrammable **ROM**) chips are often part of the changing hardware in editing systems.

FANS
When electricity flows through computer chips, they heat up. If they get too hot they can stop functioning. Consequently, all computerized equipment has a complex method for self-cooling. Fans. All computers have fans placed near sensitive locations in the computer's hardware. Without them, systems would overheat and stop working. Videotape equipment also usually contains computer components along with mechanical ones, consequently these machines need extensive cooling. Videotape facilities and edit bays are often carefully cooled for this reason.

SAVE and AUTO-SAVE
Since RAM is a temporary work space, and it is so vulnerable to power loss, computers have means for automatically saving the RAM on the hard

disk. When you select "save" on a computer, this is what is happening. The material is still in RAM, but a duplicate has been copied onto the hard disk. Most systems also have an "auto-save" function that will periodically save the data in your RAM for you, in case of accident. This way, the only material that can be lost in a power outage is the newest material edited, after the last "save." You would want a computer to save often, but since it does take some time (albeit only seconds, usually), you probably wouldn't want to do an auto-save after every edit. Typical systems auto-save every 15 minutes or every 10 edits, or some such factor. *Auto-saving is user-definable.* If you want it done more or less often, you can usually set it. Whether you are editing or doing any important work on a computer, it is important to *save* often. It only takes losing your data once for that lesson to be hard learned.

BACKUPS

RAM is only temporary memory; hard disks are "semi-permanent" — and generally considered *VERY* safe for the storage of information. But accidents do happen. Some of these accidents will be computer-related: strange power surges, "glitches" on the magnetic media, mechanical problems with the hard disk's "read" head. . . and some of these accidents will be human-error: *"Whoops! You didn't want to delete Act II?"*

For these and other reasons, it is always a good idea to have a copy of your work kept separate from the computer, just in case. This copy is called a *backup*. While some editors demand actual hard copies (printouts or videotape assemblies) of all their work, these are generally not necessary — and certainly not worth the time is would take to make them every day. The most common backup strategy is by using floppy disk (for small things like edit information) and digital tape (for large volumes like digital source).

Every day, at the end of the day, an editor may want to *copy* the entire project (the data, but not the digital source material) onto a floppy disk for safekeeping. If this is done, an accident's severity would be limited. The degree of safety you need beyond this is very subjective. On some projects, a separate floppy disk is kept for each day of the week (and one for "lunchtime"): on Wednesday you back up the "Wednesday" floppy, and don't re-use it until the next week. This, along with a single disk for lunchtime backups, gives you pretty complete security. Most users do not desire this degree of coverage: it is considered satisfactory to back up on a single floppy twice a day — at lunch, and at the close of the edit session.

On systems with too much data to fit on a floppy, small *digital tapes* (like DAT or Exabyte) that hold large quantities of information can be used.

It is actually pretty rare for a serious malfunction in an editing system to lose an editor's work; but on those odd occasions, the backup is priceless.

HIERARCHICAL FILE SYSTEMS

Computers, and specifically editing systems are based on hierarchical file systems, or HFS. Although it sounds more confusing than it is, an "HFS" simply means that there are multiple layers of "depth" into which files can be organized.

The most common metaphor for a computer's hierarchical file system is simply a filing cabinet. Everyone has seen a filing cabinet:

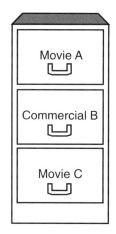

When you first walk up to the cabinet there are drawers in it, and each drawer has a label on the front. In editing systems, this is the directory where all your projects will be named . . . You can peruse all the choices, open up one, look around, close it up and choose another. But let's say you choose one. You open it up. You are now inside the project.

What do you see inside? All projects are essentially two things: big files of digital audio/video material, and small files made up of pointers to pieces of the digital footage. Another way to view these is as *source material* and *edited sequences*.

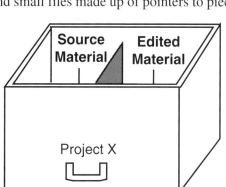

Source "dailies" are going to be presented to you in some sort of special location — a bin or a listing of shots or perhaps just the picture labels for a series of shots. Edited sequences, and there might be many of them, will also probably be together and separated from the dailies, but certainly not always. You can look over all the choices you have edited previously, pick one and re-edit it, or you can create a new place to build a new sequence. These are the kinds of things you can do once you are inside the project or production.

If you select one of your edited sequences, and *open* it up to begin re-editing, it is as if you chose a file in this file drawer and pulled it out and opened it up. You are now 2 levels into your editing system — first you opened a project (like opening a file drawer), then you opened up an edited sequence (like opening up a single file), and now you are working there. Since a computer is often like a very organized person, you often must close one file before opening another. And you must close one project before opening another. Not only does this keep things neat, but it sometimes is the only way in which a computer can work.

Most systems have subtle variations on the HFS, and usually name the file drawers and files differently (a Macintosh has "folders" and "documents"). But the idea is pretty consistent from system to system. Familiarity with personal computers (like the Macintosh Desktop, or the IBM-familiar file tree) helps a great deal in learning and understanding the organization of a computerized editing system, although it is by no means a prerequisite.

RELATIONAL DATABASES

One of the advantages computers have in the editing world is their fantastic ability to organize and manipulate data. Editing is a data-intensive exercise. For videotape, data include computer-read timecodes, record points, list tracing, and so on. For film editing, there are ample key numbers and code numbers that must be logged and tracked. Code numbers are linked to both their original key numbers and the scene numbers of the dailies. Reels are built and re-built, and scenes are moved around. Bins of trims hang in every corner of the rooms; although the editing itself is not number intensive, locating shots that have been cut and moved and hung up and changed . . . without losing them . . . is an ordeal fit for any computer.

A *database* is a special file where discrete pieces of information are linked up, and made accessible. A Rolodex is a kind of database. Information of all types can be stored on "cards" that have on them any number of smaller pieces of information, called "fields." The *fields* in your average Rolodex might be "name, address, or phone number." The fields in your average film log might be "scene, take, description, day shot, film roll number, lab roll number, key number, or code number." Traditionally, film assistants build different books of related information, but until recently they have had to do this entirely by hand. Now computers can be used to make this organization easier.

Most nonlinear systems have a "log," which is a database of information about all the source material. For a given strip of film, a database might include any or all of the following:

⇨ Scene/take numbers

⇨ Shot description and/or key words for search fields

⇨ Start film key numbers and/or code numbers

⇨ Associated camera roll, lab roll, and sound roll numbers

⇨ Nagra timecode number

⇨ Source videotape (after telecine) timecode number AND 3:2 pulldown sequence information

⇨ Date shot

⇨ Film type (35mm, 16mm, 3-perf, 4-perf, 70mm, Vistavision, *etc.*)

⇨ Film frame rate (24fps, 25fps, 30fps, *etc.*)

Although an editor might only need a few of these pieces of information, for instance, the scene/take number, the database contains *all* pertinent information. Data like *shot duration* can always be calculated from database information

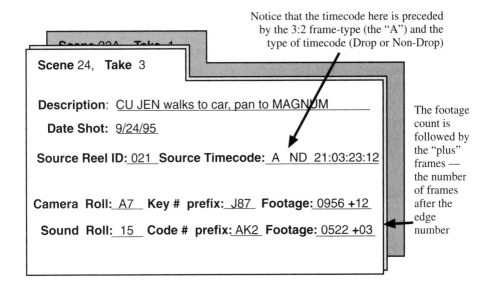

Notice that the timecode here is preceded by the 3:2 frame-type (the "A") and the type of timecode (Drop or Non-Drop)

Scene 24, Take 3

Description: CU JEN walks to car, pan to MAGNUM

Date Shot: 9/24/95

Source Reel ID: 021 **Source Timecode:** A ND 21:03:23:12

Camera Roll: A7 **Key # prefix:** J87 **Footage:** 0956 +12

Sound Roll: 15 **Code # prefix:** AK2 **Footage:** 0522 +03

The footage count is followed by the "plus" frames — the number of frames after the edge number

One feature of most "film-style" systems is that database information is generally "buried" in the system, so that the editor does not have to have much contact with numbers and data that are unnecessary or even a hindrance in the creative process. Systems do not always have a way to display film numbers and other filmic data. Some have virtually no displayed numbers at all, short of the scene/take labels.

Unlike a Rolodex, nonlinear editing systems have a *database management system* (DBMS), which involves the complex sorting and tracking of information about the original pieces of film and the tape transfer. How much manipulation you can do varies a great deal from system to system. Some systems allow virtually no sorting, searching, or re-organizing of the data in the database. Others have extremely high power to add fields, sort, print, and modify the human interface to the database. Actually, for most projects, there is usually little need to do very much modification to general database functions, but in each case, you want the manipulation powers you have to be based on the kind of project you are doing.

A *relational database* is a specific subclass of index that you might need. It is often manifested in an editing system's *source log*, that relates both timecodes (from the telecine transfer) and film edge numbers. By interlocking these pieces of information, along with 3:2 type, a computer can translate timecodes into film numbers, or vice versa.

The equivalence between videotape and film is "locked up" at "index frames." These frames are designated in telecine usually by a hole-punch made at a key-numbered film frame. By clearly marking this index frame, you can locate it in video, and log it into the database. The numbers from this point, continuing forward in the source, will be tracked. KeyKode and KeyKode readers in telecine greatly facilitate this process.

Index frames are usually the first frame of a scene, or perhaps on the clap board's sync point. As long as this frame can be easily identified, and falls near the head of a shot before which no material will be edited, any might do.

Since most systems edit and count at 30fps, they must contain a mathematical formula, or *algorithm*, for converting the videotape timecodes into related film numbers. This is done using the database. Systems that digitizes, edits and counts at 24fps does not need an algorithm to convert an edited sequence to edge numbers, however it will still need one to generate (30fps) timecodes. Regardless, all systems that output film information and tape information have some kind of relational database and internal algorithm.

Systems without a relational database often use 3rd party software products that do this same thing. There are many programs available to convert timecodes to film numbers — an early product was Adcom's *Transform LM*. Today there are others; noteworthy among them is Adelaide Works Inc.'s *OSC/R* (pronounced "oscar") — Offline Support for Conforming and Re-formatting — an IBM PC-based program for managing film and tape data.

EDITING PRIMITIVES

Let's look at a piece of 35mm film, 4 perforations per frame. The beginning of the film is at the top, the end of the film is at the bottom.

Rather than draw the film up and down, we can lay it down on its side, like this, with the beginning (or HEAD) on the right and end (or TAIL) on the left:

Actually, it really doesn't matter where you place the HEAD or TAIL of a piece of film; except that film editors are familiar with *heads* of a reel being on the right — on a flatbed, the "supply" roll of film is mounted on the left plates, threaded through the prism heads, and the "take up" reels are on the right:

TAKE UP

SUPPLY

Technically speaking, videotape runs exactly the same way: a cassette tape has the take-up reel on the right. Still, when not confronted with the reality of film or tape, many people feel that *just the opposite is happening.*

Because of this point-of-view confusion, and because there currently is no consensus, editing systems tend to locate **the head of a shot on the left** — as if you were reading the English language, and *not* like you are working on a flatbed. Since most systems display film this way, this book will do the same, although it is important to recognize that this is not universal or necessary.

The head/tail issue surfaces at two points of the electronic nonlinear interface: 1) editing transitions and 2) working with timelines.

HEAD ← TAIL

Of course nonlinear editing systems do not actually CUT or TAPE anything, just like linear electronic systems,. Linear editing is a process of selectively recording chosen pieces of source material from one videotape to another. If you want to make one shot longer, you need to go back to the source tape and record the desired shot longer.

On the other hand, film editing involves actually cutting out pieces of film from the source dailies. If you have a 100-foot shot, and you use about 4 feet from the middle of the shot, you have two pieces of film left over — called the "head trim" and the "tail trim":

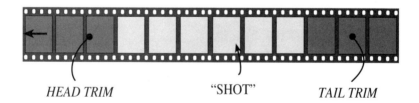

HEAD TRIM   "SHOT"   TAIL TRIM

Every time you cut a shot out from a strip of film, the remaining trims must be carefully put away. They are usually hung up together in a *trim bin* and labeled with *trim tabs*.

Electronic nonlinear editing is not like film editing nor is it like videotape editing. When you make an edit, the editing system remembers that edit location and connects it to another edit location *only in the computer's mind*. The computer uses electronic "pointers" that direct the machine to move from the end of one shot to the beginning of another. Let's examine this more closely:

FILM. Three shots are cut from dailies and taped together:

VIDEO. Three shots are recorded from source tapes onto a separate videotape master; no splices are visible:

NONLINEAR. Three shots are connected with pointers and are not actually recorded in physical form:

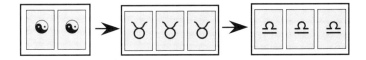

To extend a shot *on film*, you rip off the splice tape, go back to the head or tail trim to find more film, cut it in, and re-tape. All the following shots push down.

To extend a shot *on tape*, you re-record the shot you want to change, making it longer, and then tell the computer to re-record all the following shots after it.

To extend a shot *nonlinearly*, you go to the "splice" you want to adjust, tell the system to allow you to trim it, and you add the new material. The computer re-connects the pointer from the new end to the next shot. Nothing else has changed:

Since the material you see edited together is only a virtual (simulation) of actually connected shots, and since there is no "master" tape *per se* (only source media are used to simulate a master), *THE FULL LENGTH OF EACH SHOT IS ALWAYS AVAILABLE AT EVERY SPLICE POINT* without your ever needing to go back to the actual dailies (or head/tail trims).

One way to visualize this is the *folded paper method*. Think of a piece of film as a piece of paper:

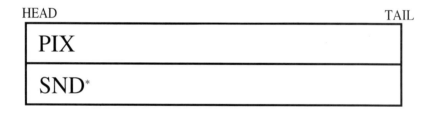

At one end, you cut the head, and further down, you cut the tail:

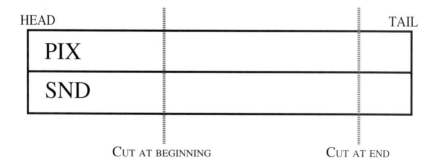

*SND is a common abbreviation for "sound"

Rather than "hang up" the head trim and tail trim in a film bin, simply fold them back:

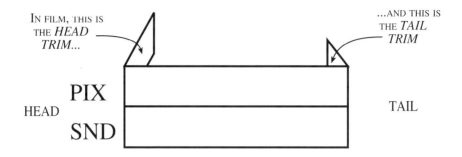

And two shots, before splicing, might look like this:

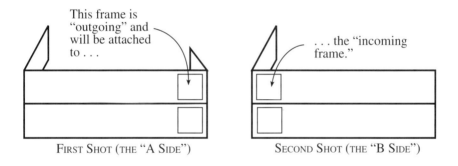

FIRST SHOT (THE "A SIDE")   SECOND SHOT (THE "B SIDE")

TRIMS

ALL nonlinear systems allow you to modify transitions between two shots already edited into the "master." Modifying these transitions involves **first** going to the desired transition and "trimming" it. Although names vary from system to system, *both shortening and extending shots are done in this way.* The idea that shortening and lengthening can be done the exact same way — and that both might be called "trimming" even though lengthening a shot seems counterintuitive to this name — can often be confusing to first-time nonlinear editors familiar with film trim bins. Video editors, though, are already quite used to the term in this context; in the video domain, to "trim" means simply to "modify" a given transition.

Virtually all nonlinear systems have two KINDS of **transitional editing** (we will call it "trimming" for simplicity here). One kind of *TRIM* allows the editor to adjust each side of the cut — the outgoing and incoming sides — separately. This is a "film-style" trim. The second modifies both shots simultaneously — for every frame that is added to one side, the exact same number of frames is lost from the other side. This is a "video-style" trim.

Here are two frames on either side of a transition — the last *outgoing* frame is shown below on the left, and the first *incoming* frame is on the right. These are usually displayed on two side-by-side monitors or digital windows:

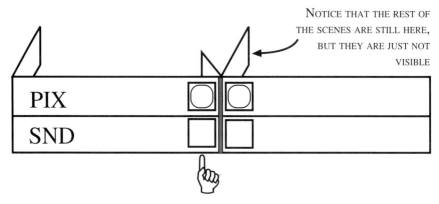

NOTICE THAT THE REST OF THE SCENES ARE STILL HERE, BUT THEY ARE JUST NOT VISIBLE

PIX

SND

Using an arrow or dot or light or some kind of pointer, you choose which side of the transition you want to adjust. All systems give you some way to select which side you want to control. Here we are indicating the A-side of the transition — the tail of the outgoing shot

A Film-Style Trim — known by many names (non-sync trim, a "yellow" trim, *etc.*) — will change the absolute length of your program, either longer or shorter. Regardless of which, the film-style trim will shove down or pull up all following edits, just as it if you were actually working on film. In videotape editing terminology, the entire list would be "rippled."

There is a non-rippling variation on the film-style trim that is much less common in nonlinear systems. If you shorten a shot and choose not to ripple the following events, the system will automatically slug in *leader* to fill the space created. This allows film-style editing to have some of the video-style (non-rippling) functionality — a particularly helpful feature in the editing of picture-only or sound-only, where you do not want to throw all following shots out of sync.

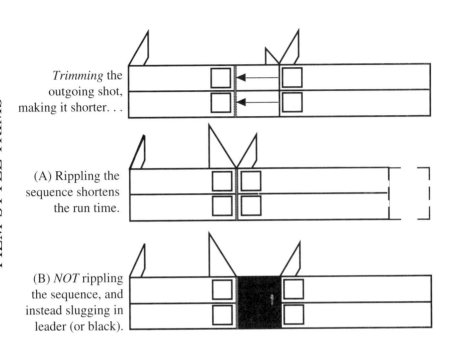

FILM-STYLE TRIMS

Trimming the outgoing shot, making it shorter. . .

(A) Rippling the sequence shortens the run time.

(B) *NOT* rippling the sequence, and instead slugging in leader (or black).

VIDEO-STYLE TRIMS

The Video-Style Trim (sometimes called a synchronizer trim or slide-cut, among other names) modifies **both** outgoing and incoming shots *simultaneously* — for every frame that is added to one side, the exact same number of frames is lost from the other side. On film, it is as if both shots were interlocked together in a synchronizer. This way, there will be no durational change in the length of the show; in video terminology, no following record points are moved, nothing is rippled. (Rather than selecting the outgoing or incoming shot, you needn't designate either side of the transition since both will be modified — or you might just designate the splice point itself.)

Notice how the incoming shot has been extended the same length that the outgoing shot has been shortened. The length of both shots together has remained unchanged.

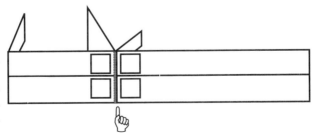

Transition editing — being able to see the outgoing frames of one shot and the incoming frames of the next shot, in order to decide upon the edit location — is an important feature of nonlinear systems. Some systems are *"TRANSITION-BASED"* in that all editing is performed in this way. Other systems are *"CLIP-BASED,"* meaning that you must decide upon a shot's in-point and out-point before splicing it into the master.

Clip-based editing is a direct descendent of video-style editing, in which you must choose a source in and source out before committing to an edit. Although the "commitment" in nonlinear editing is less severe than in tape or film editing (you can always easily change the out-point if you end up disliking it), it is a style that can be a little more frustrating to the seasoned film editor. Many times you do not want to decide on the end of a shot until you have the previous edit working to satisfaction.

While most systems have transitional-type editing standard in their RE-EDITING modes, fewer systems are transition-based in first cut situations.

SPLIT EDITS

When you trim *just* the picture without changing sound, or vice versa, you will end up with transitions in picture and sound at different locations. In film these are called *pre-laps*, or *overlaps*, or *L-cuts* (named for the shape they form). In videotape, they are *split edits* — where sync picture and sound are split apart at a transition, but continue and will return to sync later in the shot. Here is a split edit, with picture leading sound:

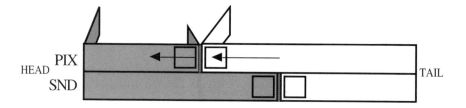

Although there are many ways to create these overlaps, the easiest for an editor to perform is a video-style (or synchronizer) trim of the picture alone, without modifying any sound. As you lengthen or shorten one side of the transition, the other side will be adjusted the same amount in the same direction. This way, all the following edits in your sequence will never be pushed out of sync. A *film*-style trim of picture-only (or sound-only) will usually create a situation where the picture runs a different length than the sound.

Because of the way shots are created and modified, students of nonlinear are often directed to create split edits by first making a straight cut (in picture and sound), and then modifying the picture. Although this is not a rule, cutting for sound first, and the overlapping picture later, is the most common and perhaps the easiest method of making overlaps.

INSERTING

If "trimming" is defined as going to an existing splice and modifying either the outgoing or incoming shot (or both), then INSERTING can be defined as going to source footage, finding a new shot, and placing it somewhere into the already-cut master.

Inserting is not necessarily *directly* from source material. Some systems have "bins" and other kinds of locations from where pieces of material can be selected for inserting. Regardless, all systems allow you to INSERT new material into the edited master.

Inserts, like trims, must be designated as "film-style" or "video-style." A film-style insert will place in the new shot, and shove all following material down in time, like in film. The master is automatically rippled. A video-style insert rolls over the existing master, obscuring old shots with the new one. If 3 feet of film (2 seconds) are inserted, then 3 feet of existing-master is taken out. Durationally, you have not changed the timing of the program.

Some systems see the video-style insert as somewhat *dangerous* (a "red" insert) because you are always going to lose existing master material, and if you insert too much, you might "record over" something you wanted. Thus, it can be risky. Other systems ironically see video-style inserts as the *most safe* (a "green" insert), as there is absolutely no way to throw a sequence out of sync by performing an insert of this kind.

On systems where losing sync between picture and sound tracks is risky, easy to do, or hard to fix, film-style trims and inserts can be dangerous. Video-style editing, then, is seen as safe.

Film-style inserts can also be characterized as "4-point" edits. For these, an in- and out-point must be determined on the new shot as well as in

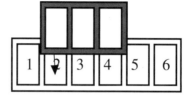

Here a new shot (in grey) is being inserted into the master. . .

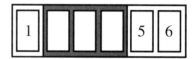

A *film-style* insert will break open the master and drop in the new shot — shoving the rest of the sequence down in time.

*Note that this insert "steps on" the negative cut margin.

A *video-style* insert places every frame of the new shot *over* a frame of the old shot. The total length is unchanged.

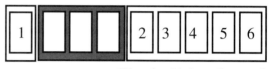

the master. In general, the computer will calculate the fourth point if the editor designates any three.

By the same logic, video-style inserts are sometimes described as "3-point" edits. Here, an editor either chooses the exact dimensions of the new shot (an in- and out-point) plus where the shot should go in the master — or vice versa (the exact location is specified, and a beginning or end point is chosen in the new shot). Either way, 3-point edits can "roll in" or be "back-timed" over existing material in the master.

LIFTS

Many nonlinear systems offer another way to shorten the edited master. You can "lift" a frame, a scene, or an entire sequence the same way: mark the first frame, then the last frame of the desired material, then "lift" it out. And like trims and inserts, a lift can be film-style — where the following shots are rippled; or video-style — where leader is slugged in the place of the lifted shot, thus not affecting the timing or positions of any other shots.

Once the material is lifted, exactly where it goes is variable:

☞ The LIFT can go to a special "bin" where it is stored until later use. These bins hold a finite number of lifts, but it is usually a large number.

☞ The LIFT can be "thrown out." Since there is no concept of "head trims" and "tail trims" as there is in film editing, there is no need to put a piece of footage away for later use. The source material is itself never changed or broken down while editing. No film is ever really *deleted*.

☞ The LIFT can be moved to another location in the master.

With regard to this final type of lift, it also seems to define a certain subset of nonlinear systems. Clip-based re-editing (unlike transition-based re-editing) might involve re-arranging existing shots. For example, if a sequence is assembled as shots 1, 2, and 3, you might want to switch the order of the last two:

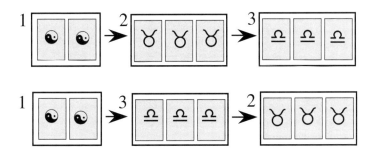

Although this function is impressive, it is often more showy than useful in editing narrative material. It is best for manipulating the order of shots or sequences in demo reels, slide-type presentations, and some kinds of documentaries. Most digital-based systems do this easily, while tape-based and disc-based systems will be somewhat slower at the same functions.

For clip-based lifting, you simply need to choose which shot or shots you want to move and point out where you want them to go. For moving material not already defined by shot boundaries, the head and tail locations of a lift must be marked, then LIFT selected. This material will be removed from the master.

RIPPLE

Although differentiating between film-style and video-style for trims, inserts, and lifts would seem to be sufficient, there are a few cases of the film-style edit that require an additional decision. "Rippling" is actually a specific video term for adjusting subsequent EDL events when shots are added or subtracted — and thus the record times are changed.

For nonlinear editing, although there are no event numbers *per se*, "rippling" is simply the unlocking of shots to time —past a given spice point. By definition, video-style edits do not involve rippling. But film-style edits that shorten master durations might ripple or not ripple (the alternative to rippling is slugging with leader to maintain sync or time).

Virtually all systems have methods for performing these functions; they are fundamental to mastering editing:

Trims and inserts and lifts.
Film-style and video-style.
Ripple (on or off).

Combine with these functions the ability to edit picture and sound together or separately, and you have a number of complex editing concepts that can be simply represented with very few variables (and thus, few buttons).

SPLICE

It might seem extraordinarily fundamental to be able to "splice" two pieces of film together, but some systems do not have any way to do this. On these systems, all first cuts are built using "inserts." Still, both clip-based and transition-based systems usually have some central button labeled "splice" or "join," which will attach a new shot (as from the dailies) to some shot in the edited sequence. For clip-based editing systems, like traditional video-tape editing, you will have determined both an in- and out-point and then splice the entire shot into your master. For transition-based systems, you will be attaching the last frame of one shot to the first of a new shot, with no other required decisions. In a computer sense, the splice button is a kind of "ENTER" function (like pressing "return") that commits your edit decision (whether it be a transition re-edit, an insert, or something else) to the memory of the system.

UNDO/CANCEL

Most computers are able to undo the last function performed. Electronic editing systems are no different. Virtually all systems have some kind of UNDO or CANCEL button. If you make an edit you don't like — whether it be by accident or on purpose — you can press this button and return your edited master to its earlier condition. Different systems have their own unique methodologies for canceling mistakes or going back to prior states.

A few systems allow for canceling a number of events — sometimes as many as 32 — by repeatedly selecting the cancel function. On the one hand this can be extremely powerful, allowing an editor to work backward through a troubled edit list, edit by edit, until the source of the problem is undone.

On the other hand, it can be more confusing than not. In a few systems there is no way to know what each selection of CANCEL is canceling, so it can undo work without the editor knowing exactly what is being changed. Also, trying to undo a single edit might involve removing work that an editor wants to keep. In this way, multi-layered canceling is powerful but potentially more trouble than it may be worth. Still, a few levels of cancel is a very useful feature.

NUMERIC EDITS

On editing systems that allow transitions to be adjusted numerically, using timecode or just numeric values, lengthening or shortening shots can be a little confusing. Here is a shot from our EDL:

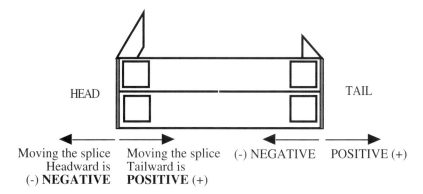

If we wanted to make the shot 1 frame LONGER, we could specify the tail of the shot, and "trim the out-point 1 frame" which means we would add one (+1) frame at the "out" timecode. Similarly, to shorten this shot 1 frame we would trim 1 frame from the out point (or "-1" at the tail). This is logical if you think of the "1" as the value, and the -/+ as the direction.

Mistakes are often made because people **incorrectly** interpret the syntax to mean "you lengthen a shot by ADDING (+) frames and shortening a shot by SUBTRACTING (-) frames." *This is not true.*

At the head, to make the shot one frame longer, you would want to start the timecode of the "in" one frame earlier, in other words trim minus one frame (-1 frame). In the same way, to shorten the shot, you would want to trim plus one frame. What this means (look at the diagram) is that to move an edit point to the *LEFT* — at both the head or tail — you use a **minus** sign. To move an edit point to the *RIGHT* you use a **plus** sign.

This is a fundamental concept in linear videotape editing, but is often confusing to film editors attempting electronic editing for the first time. Nonlinear systems generally allow for visual and physical modification of shots *in addition* to numeric trims, and rarely rely on the typing of numbers as the only way to modify edits.

EFFECTS

Most, if not all nonlinear systems have at least some video effects capabilities. The most common, and arguably the most necessary, is the ability to preview fades and dissolves.

Not all editing systems offer effects capabilities in their basic system — but only as an upgrade.

On many digital systems dissolves and fades as well as other video effects must be pre-built automatically by the system ("rendered") as separate elements (like film opticals), as opposed to actually triggering the effect live each time the transition is presented ("real-time").

Motion effects are available on some systems. These include speed-up and slow-down of shots. More sophisticated effects involve stretching a shot to fill a desired space durationally. Although powerful video tools, these effects can be difficult to re-create in film.

Often available are some kinds of text entry: sometime these are quite simple, but other times these rival professional character generation tools.

Systems that do not actually preview effects are still capable of including the information about each effect in the video EDL: effect type, transition rate, wipe code (if necessary) will be included for online. Much rarer is the translation of these video effects into optical specification print-outs that can be delivered to film optical houses for the re-creation of video effects using film elements.

For effects created in video, there *are* methods available to convert video to film opticals — although they are somewhat expensive and are generally priced by effect duration. Care must be taken in the offline of projects with both film and video finishes, not to create unique online effects that must then be matched with film opticals. Today, more and more film effects are being created in the digital domain, on dedicated high-end computer graphics (CG) systems.

MONITORS

The placement and useage of video monitors was at one time a somewhat complex thing — and seemed to vary with virtually every video and nonlinear system. Systems generally had a two-monitor set up: source on one side and master on the other. In many cases these two video monitors were not sufficient, and a third monitor was required exclusively for the editing data.

Most commonly, source material was shown on the right and the edited material (master) was shown on the left.

There is another use of a two-monitor juxtaposition when editing, and that is for *transition editing*: that is, when the editor can simultaneously view the outgoing frame of one shot and the attached incoming frame on the other. This feature was not common to all early nonlinear systems, but when it was available, it often used the source and master monitors for the outgoing and incoming frames.

Confusion with the layout of video monitors came from a number of situations. In the first place, film editors who worked on flatbed tables were used to source rolls being loaded up on the right side of the table and the take up reels on the left. This led to problems in transition editing because in a film editor's mind, the head of a shot was on the right and the tail on the left. Video editors, when working with transition editing, and not used to having a strip of film before them, tended to visualize the film running left to right, the way one reads english, and wanted monitors to represent this. Consequently, film editors and video editors moving into nonlinear editing would request different monitor functionality and configuration.

While viewing all source and master material in a full-screen mode seemed ideal from an aesthetic viewpoint, it could be logically difficult. Digital systems solved many of the confusing monitor layouts simply by working with smaller images. Rather than use banks of monitors for views of multiple shots (whether in a true "multicamera mode" or simply viewing different takes), digital systems used "windows" imbedded in the display to show smaller views of shots.

Digital systems tend to have at least one, and often two graphics monitors, neither of which are specifically "source" or "master." Images are presented in windows that can take on many sizes and can be located in either monitor. Typically, there is a central editing monitor upon which everything is or at least can be accessed: timelines, source clips, transitions, and so on. But managing all these overlapping windows and many source clips can be daunting, and a clear advantage is reaped when two monitors are used in tandem. In the traditional two-monitor set up, the monitor on the right provides a location for spreading out small *thumbnail* images of each source take. The left monitor is the editing and system environment, and will have

imbedded in it locations for "source" and "master," equivalent to the stand-alone video monitors used in the earlier systems

When involved in transition editing, the master window itself is often subdivided to show outgoing and incoming frames; although it is often possible to transfer those tiny frames to the full screens of the two system monitors.

When digital systems use "multisync" monitors, it means that the monitor can present either the computer data, or actual video (like a TV set). If both monitors are multisync, it might imply that each is interchangeable in functionality, however software and cabling usually limit this.

Many systems have interface styles that are configuarable (and thus personally designable) as to the locations of windows and functions, although some editors prefer a rigid window format that makes learning the system consistent and straightforward. Whether on the left or right, full screen or thumbnails, configurable or rigid, editors easily adapt to the given locations of the elements of the editing system.

T I M E L I N E S

Editing systems that provide some sort of graphic display often use a device called a *timeline*. Although timelines in various forms have been used for eons, their application in nonlinear editing, specifically in making computerized film-style editing more accessible, was first introduced on the EditDroid in 1984.

Since then most systems have adopted similar or their own versions of the timeline. Here is a simple example:

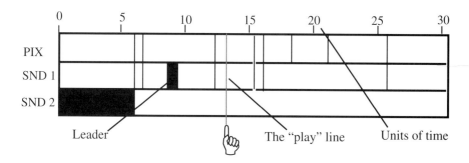

Although horizontal timelines are not necessarily the most easily manipulated shape, virtually all system depict them as running from side to side on the editing screen. If the timeline were designed to depict a flatbed, time zero ("heads," or the beginning of a sequence), would start on the far right-hand side. Unfortunately for film editors, virtually all systems provide for *time zero* to be on the leftmost side ("heads" on the left). This would be consistent with the transitional edit method of having outgoing A-side frames on the left, and incoming B-side frames on the right. Whichever method a system uses, the timeline should be consistent with the transitional logic.

A timeline is actually a series of "tracks" running together as if in a synchronizer. Picture, usually on the top, may or may not be labeled as such. Audio channels lie underneath. Newer systems offer many more (sometimes unlimited) "virtual" tracks of audio, but may be limited to playing only 2 or perhaps 4 simultaneously. Also if the system has special effects tracks, (for keys, titles, *etc.*) they tend to be shown as running above the video.

In audio, another feature is growing more prevalent. Rather than simply show a graphic bar representing sound, actual waveforms of the sound can be displayed. These audio waves, made possible by the system's use of digitized audio, allow for accurate visual guides to individual parts of sounds. One of the earliest uses of this display was in SoundDroid prototype, from 1985.

The existence of a timeline offers at times a blessing and at others a curse. The blessing is obvious: editing electronically has removed the last vestiges of physicality in editing — there is nothing to hold onto, no edits actually taking place, nothing being cut. A timeline provides a physical object to which editors can relate their work. As well as appearing as a visual guide, the timeline makes computerized editing a more geometric exercise — like film — in that you are moving shapes and changing them until they fit in appropriate ways. It usually makes the computers easier to use.

The curse is less tangible. A timeline, like the icon, is still just a computer phenomenon. Having a timeline generally means that an editor is dependent on it for the execution of edits. Contradictory to many product claims, active timelines mean that the computer is not in the background, but *central* to the editing process. You must "point and click" on the display to perform certain editing functions. Regardless, all systems that have timelines are considered simpler to use than systems without them.

Aside from where heads and tails are found on a timeline, there are a number of features that are associated with this kind of graphical display:

• ZOOMING IN AND OUT. Although the horizontal or vertical displays of the timeline are bounded by the screen, the timeline in its entirety may be enormous. Consequently, users often want to scale the display in various ways. The smallest functional unit of picture editing is the frame; this is the closest users would ever need to expand. By zooming in, very short edits can be easily identified and manipulated, where they wouldn't be from *farther out*. With point-and-click editing, zooming in (and out) is essential to locate relatively short shots. The "farthest" unit an editor might want to see would be the sequence in its entirety — 30 seconds, 10 minutes, 2 hours. . . .

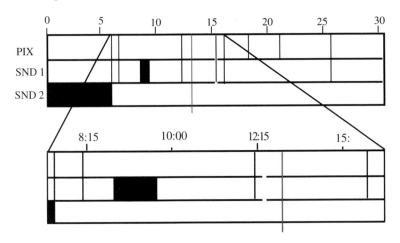

Most systems with timelines have the scale user-configurable, but this is not always the case.

• SCALE UNITS: Although most electronic editing systems arc based on video timecodes, some allow for the scale to be converted to feet and frames. Record times (run times) are usually displayed along the axis of the timeline, allowing for quick identification of relative lengths of shots, and approximate durations in a cut sequence.

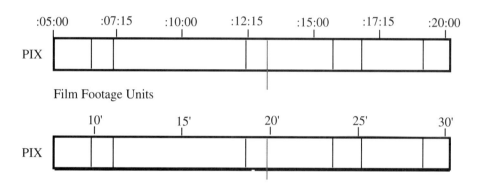

• IDENTIFICATION: A graphic display is just that — graphic. In addition, some method for identifying the source information for a single shot is usually provided somewhere on the display. Text (scene/take information, source in/out points, descriptions, video event numbers) can be located in affiliated displays, or if room is available, actually inscribed in the timeline itself. In either case, this information is important to have in order to use the timeline effectively.

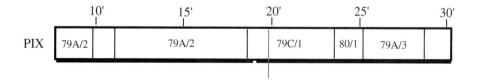

• COLOR: Sometimes color or shading is employed to help distinguish shots originating from different source dailies. This is an ideal way to

identify the relationships between picture and sound, especially in complicated editing where there are stolen lines, split-edits, and frequent differences between source picture and track.

• DURATIONAL ACCURACY: Older systems sometimes had timelines that
uniformly represented each edit, regardless of its actual duration. Today, all
timelines represent shots with lengths that are proportional to the shot's
duration.

• CUE POINTS AND MARKS: A relatively new feature allows an editor to
place special marks in the timeline for quick cueing. While the number or
available marks make be limited, they are sometimes essential to syncing
material up or locating chosen locations in long shots and sequences.

• "NOW" LINE OR "PLAY" LINE: There is always some indicator on the
timeline of where the virtual master is currently positioned. The timeline
will either move past a fixed "head" location (like celluloid on a film table),
or the "now-line" will run back and forth over the timeline. The moving
now-line is the most common paradigm.

A secondary but important feature of the now-line is its ability to track
the rolling master wherever it plays. Some systems only update the timeline
when the master stops playing; more better systems actually scroll the
timeline or now line as the master plays. This feature is beneficial to editors,
but does involve more computer work (screen refreshing). Also, moving the
now line rather than the timeline will force the timeline to "jump" when the
now line reaches the edge of the screen — and can in some cases be
confusing.

• VARIATIONS: Although most timelines run horizontally, some systems
have them running vertically. The traditional audio "cue sheet" is a vertical
timeline — as are many digital audio workstation displays that mimic these.

Also, many timelines can be converted to unique presentations of the
edited sequence, among them "head views" and "head-tail" views — that
show small images of the actual frames at the splices, adding to the graphic
nature of the computerized editing system.

In whatever format, timelines often provide for simple moving of shots,
re-ordering existing sequences, and easy deletion of unwanted shots. Each
system's functionality regarding timelines should be investigated.

Clear and easy-to-follow timelines are a real asset to a nonlinear editing
system.

D I G I T I Z I N G

With the generation of digital offline systems comes the art and science of digitizing compressed video. This is the process whereby the source video is converted from NTSC (or PAL) video into digital data, and stored on some type of magnetic storage media — usually hard disks, but occasionally, magneto-optical disks.

Digitizing is the process that precedes all other editing activities, and may be separate or part of the "logging" of the source material. At a purely technical level, digitizing begins with the feeding the video source into a digitizing card that is connected to the computer; the card samples the video, assigns numeric values to each pixel, and records them onto a disk.

Video is traditionally digitized via an **Assembly** Method; less common is the **Dub** Method. Both will be described here.

The Assembly method involves creating a type of off-offline edit with a videotape deck and an editing system (or a remote computer). The editor or assistant rolls through a videotape marking in and out points of the desired shots on the tape. This creates a *kind* of videotape EDL, but for raw source footage and associated data. This special decision list is then used by the editing system to perform a *Batch Digitize*. Batch Digitize is an automated process for digitizing each shot as designated on this EDL. It is a sort of auto-assembly, complete with an edit session, a source tape, and a master (in this case, a digital random access master). Batch Digitizing allows material to be logged remotely, without tying up the editing system with the somewhat laborious logging session — you know: selecting of the heads and tails of each source clip. It is done with increasing frequency as the cost of the digital editing system increases in cost.

If tying up an editing system is not too costly, an editor may use the system as a tape controller and mark ins and outs, then record (*i.e.* digitize), each desired shot. Regardless of whether digitization is done immediately, or later, this method requires the pre-selection of the desired source material. If source material originated on film, and was then telecined to videotape, often the "EDL" for assembling comes as a Telecine Log. This telecine EDL provides the data for the auto-digitization of the source clips. When there is a large amount of source material relative to selected shots, the assembly method is most efficient.

The Dub method requires the dubbing of the entire source videotape into the digital editing media. While this method allows for the immediate availability of all source material, it is generally considered to waste a large amount of valuable digital disk space with obviously extraneous material. Still, it removes the necessity of the "off-offline" edit (the laborious pre-selection of shots). The Dub method is inefficient except when it is exceptionally easy to discard extraneous pieces of source material at will. With this functionality, source material is removed from the digital storage device as it is determined un-useful rather than selected from the videotape prior to editing. The dub method is relatively uncommon, especially as most systems cannot delete the source data for only a portion of a digitized clip.

Digitizing is in real time and always uses a video compression scheme to reduce the huge amount of data in the video frame into a usable and manageable size. Most compression rates today are around 20:1 (common in JPEG compression for offline), but for high-quality video, the rate approaches 8:1 or more. Editors need to determine the degree of compression required prior to the starting a project. This compression factor is decided based on estimates of the total required source material, and/or the amount of digital storage available at a given cost, and/or the desired resolution. Because of the popularity of the Avid, its "Avid Video Resolution" or **AVR** number, has come to be a kind of benchmark for comparing resolution quality. AVR numbers range from 1 (low) to 6 (high), but with "online" qualities being shown with 25, 26, 27 (representing 2-field digitizing at an AVR 5, 6 or 7-type of quality). Avid also offers "enhanced" resolutions that squeeze about 5 to 20% more material into the same space at the expense of a slight increase in picture "noise".

Audio is generally not compressed when digitized. CD-quality audio is 16-bit sound, sampled at 44.1KHz. Audio is frequently digitized at this quality, allowing for "online" type audio work along with the video; however disk space can be saved by digitizing the audio at 22KHz or even 11KHz. For some specific applications, audio is digitized at 48KHz, which would of course utilize additional hard disk space.

An important distinction between systems has to do with exactly when in the process of working on a project is the digitizing resolution set. As a general rule, whatever resolution is chosen, it must be decided prior to digitizing each clip. In general, a compression rate is chosen for an entire project; and then the editor is not required (or allowed) to modify this rate for the duration of that project. For other systems, the rate is shot-specific, meaning that the rate can be changed as required on a shot-by-shot basis. Clearly, the later method provides a greater degree of flexibility with regard to editing performance.

The idea of "appropriate" image resolution is a subtle one. Every type of digital compression has associated with it some kinds of specific artifacts — visual anomalies — that are created in the images. If these artifacts are understood, certain problems can be avoided. For example, when you see demonstrations of digital compression at trade shows the demo material tends to have bold bright colors (azure blue sky at the beach, deep green water, bright yellow towels, *etc.*) and lots of close ups (eyes, faces, hands). While the average viewer might regard these as good examples of digital quality, these close shots and bold colors actually tend to mask many compression artifacts. A close shot may look like it is showing more detail, but only of the subject and not of the video. A far more difficult image would be one with lots of motion, lots of tiny details, and few patterns. A crowd in a football stadium. An autumn forest in New England. A bag of multicolored beads. A shadowy figure running through an alley. Shots that show the familiar artifacts and are good litmus tests for digital systems. If you have scene-by-scene control over the digital compression during a project, you might then decrease resolution for medium and close shots, and increase it for wide or detailed shots.

Artifacts include such problems as "scintillation" (tiny sparkles in detailed regions of the frame), "quantization noise" (the normally-stationary background elements may appear to be busy with small wandering colors), "color banding" (patterns of color that show up as bands in smooth areas), and the all-too-common "blocks" (anomolies created from the DCT mathematics that create hard edges in otherwise regular areas of the video).

Offline nonlinear systems digitize of 1 field of video for every frame — this creates images with half the effective resolution of the original videotape. In these systems most default to digitize **field 1**. This creates good transfers for video-originated source, but causes some motion artifacts for material that has been 3:2 transferred from film. The only way these motion aberrations can be avoided is to digitize at 24fps, with an understanding of the type of 3:2 pulldown used to create the videotape. Dedicated nonlinear systems designed to work with film accommodate a real 24fps digitization. This is the preferred method for digital offline of film as it removes motion artifacts and saves 20% of the disk space as needed for the same material transferred at 30fps.

VIDEOTAPE   ASSEMBLIES

The degree of nonlinearity is largely a function of how long you can edit without committing those edits to videotape; however, some projects are not entirely nonlinear. There comes a time when it is both desired and perhaps required that a linear assembly to videotape be performed. All nonlinear systems, regardless of how they achieve nonlinearity, can record edits to videotape.

There are two primary ways in which a system can do this. The first and most common is the "print to tape". Less common on current nonlinear systems is the somewhat traditional "auto assembly".

Printing-to-Tape is a real-time assembly and wholly unlike standard auto assemblies. Since digital systems can preview long portions of edited material directly from the system (often as long as an entire production), they have the ability to record it onto a "downstream" videotape. While most professional systems have machine control (to control a tape deck for digitizing), others may not. With no machine control available, a print-to-tape option is the only way to get the video edits directly out of the system; the user would need to run a cable from the video/audio outputs of the system to the inputs of a video deck of some kind. Then the user would need to manually start the recording on the deck and then start the playing of the edited sequence.

But even with machine control, there are issues of print-to-tape that involve the frame accuracy of the tape produced, especially when the project is to finish on film.

The image quality of the print-to-tape output is going to be one generation worse than the video was when it existed on the system in the first place (remember that it is a digital to analog conversion to get the output onto videotape). If image quality is an issue, some kind of tape-to-tape auto assembly must be performed.

The **traditional methods of videotape assembly** — non-real-time tape to tape recording controlled by the editing system — can often be employed to get an online master tape out of the offline editing system. To do this the editing system would need to be able to control more than one video deck simultaneously. These slow assemblies can be in either A-mode, B-mode, or C-mode, with B-mode being the most common option.

A-MODE ASSEMBLIES record each shot in the order they appear in the edit decision list (EDL). These assemblies build the tape sequentially, as it will appear when complete, with no regard for the location of the source material. A-mode assemblies are often considered inefficient, though more flexible in online sessions should changes be required.

B-MODE ASSEMBLIES record shots in the order they appear in the edit decision list, just like in A-mode, except that if a source reel is missing, the shot will be skipped. Every edit that can be recorded from available source tape(s) will be. When this *pass* is completed, other source tapes will be "requested" by the computer; once loaded, the missing shots will be assembled. If more than one source is accessible at one time, systems will use all available sources. This kind of assembly is often called a "checkerboard" assembly since it will leave black "holes" in the recorded master, to be filled in later by not-yet-loaded source tapes.

C-MODE ASSEMBLIES are very much like B-mode, except that the order of assembly is not determined by the EDL but rather the ascending scene location on the source tape (like a film pull list). This kind of assembly makes most sense when performing a tape-to-tape assembly with only a single source at a time.

As a general rule, you want to do a B-mode if the record tape is long and the sources are short (or fast); a C-mode if the record is short relative to the sources that are long (or slow).

A, B, and C-mode auto assemblies run **out of real time.** Depending on the length of the source tapes, the number of edits, and length of each edit, a videotape assembly can take anywhere from 5 to 8 times the program's real duration. Because of this, the longer a project can refrain from assembling a videotape copy, the faster editing and re-editing can progress. In fact, this is one of the attractions to nonlinear editing; frequent tape assemblies considerably reduce the cost and time efficiency. A simple tape version of a nonlinear project — for example a 1-hour program — can waste an entire day that could have been spent editing. The advantage of the print-to-tape option is clear: a real-time assembly takes considerable labor and time out of the required videotape delivery. For shorter projects the penalties of an auto-assembly are much less severe. Realize that linear systems, although they edit considerably slower and with greater difficulty, are building the videotape of the edited project as they go. Like film projects cutting film, projects cut via linear videotape systems are ready for delivery as soon as a editing is completed. Neither film nor linear videotape requires *additional* "assembly" time to view a full project.

Many of nonlinear systems utilize a hybrid approach to assemblies to gather the best of both worlds. They will record their high-quality audio directly from the system in a real-time print-to-tape. Only then, the slower B-mode/C-mode types of tape-to-tape assemblies will be performed for picture-only. This will cut B-mode assembly time dramatically, especially on projects with many audio edits, but will deliver the highest quality output from an offline system in an optimal amount of time.

PRINTOUTS

All systems have paper (or "hard copy") printouts as an output option from the computer.

The most common type is the **Edit Decision List** (EDL). You can designate a particular edited sequence and then the output format: Drop or Non-drop; CMX, GVG or other type; a record start time, and EDL file name. This is an example of a CMX-type EDL:

```
TITLE: REEL 4              sc. 20 - 27              5,6,7

FCM: NON-DROP FRAME

001    BL     V     C           00:00:00:00 00:00:00:00 01:00:00:00 01:00:00:00
001    05     V     D     030   05:27:10:09 05:27:11:09 01:00:00:00 01:00:01:00
002    05     V     C           05:27:11:09 05:27:39:15 01:00:01:00 01:00:29:06
003    05     A     C           05:26:56:27 05:27:00:17 01:00:00:00 01:00:03:20
004    BL     A2    C           00:00:00:00 00:09:45:05 01:00:00:00 01:09:45:05
005    05     A     C           05:27:13:29 05:27:39:15 01:00:03:20 01:00:29:06
006    05     B     C           05:28:04:02 05:28:12:25 01:00:29:06 01:00:37:29
007    05     V     C           05:29:01:07 05:29:08:29 01:00:37:29 01:00:45:21
008    05     A     C           05:29:01:07 05:29:05:02 01:00:37:29 01:00:41:24
009    07     A     C           07:04:07:02 07:04:23:28 01:00:41:24 01:00:58:20
010    07     V     C           07:04:10:29 07:04:19:07 01:00:45:21 01:00:53:29
011    07     V     C           07:04:18:07 07:04:18:07 01:00:52:29 01:00:52:29
011    06     V     D     060   06:05:52:12 06:05:54:12 01:00:52:29 01:00:54:29
012    06     V     C           06:05:54:12 06:05:58:03 01:00:54:29 01:00:58:20
013    06     B     C           06:09:24:27 06:09:28:04 01:00:58:20 01:01:01:27
014    06     B     C           06:07:21:24 06:07:31:09 01:01:01:27 01:01:11:12
015    06     B     C           06:04:12:06 06:04:29:05 01:01:11:12 01:01:28:11
016    06     B     C           06:06:26:13 06:06:35:17 01:01:28:11 01:01:37:15
017    06     B     C           06:10:03:07 06:10:05:19 01:01:37:15 01:01:39:27
018    06     B     C           06:11:59:19 06:12:05:12 01:01:39:27 01:01:45:20
```

Next are the **film cutting lists**. There are primarily four kinds: *assembly lists, pull lists, optical lists,* and *change lists.* For editing systems that work internally at 30fps (like video source material), they have within them algorithms for converting 30fps timecode into film numbers. For 24fps systems, the generation of film lists does not require such math. In both cases, a relational database will provide the link between the timecode numbers determined during the telecine session and the film numbers, allowing the system to "trace" (or match) back to the film. Usually this software is an integral part of the editing system, but more and more video editing systems require a separate computer program to complete these functions.

> ➤ An *assembly list* is made up of pairs of edge numbers, one for the in-point and one for the out-point of each shot in the cut sequence (like an EDL). These shots are listed in sequential order according to when they occur in the

final edited sequence. Below are are two different assembly lists:

Courtesy of LightWorks

Edit:	SC 50 VER 2			Project: MRS. DOUBTFIRE		
Edit dated:	21:08, 15 Sep 1993			Format: 35mm		
Report dated:	21:31, 16 Sep 1993			Cutting-copy version 4 (6.1)		
Assembly list						

	Start	Length	Roll	Shot name	Start code	End
1	0.00	13.14	A19	50-4	KJ458822 5117+10	5131+07
2	13.14	2.09	A19	50A-2	KJ458822 5273+03	5275+12
3	16.07	13.11	A19	50-4	KJ458822 5136+07	5150+01
4	30.02	9.10	A19	50A-2	KJ458822 5250+06	5259+15
	DISSOLVE	3.00		Mark from	5258+08	

SEQ #	LAB ROLL #	SCENE	TAKE	HEAD CODE #	I	TAIL CODE #	I	I CUMULATIVE I DURATION
001	R-15	21	17	026	90.12 I	026	134.08 I	43.13
002	R-15	21	26	026	247.11 I	026	260.12 I	56.15
003	R-15	21	33	026	388.12 I	026	400.05 I	68.09
004	R-18	22X	14	024	121.05 I	024	133.10 I	80.15
005	R-15	22B	11	028	370.15 I	028	377.15 I	88.00
006	R-16	22	22	029	231.03 I	029	236.00 I	92.14

If a film cutting list type is not specified by name, it is generally an assembly list.

➤ A *pull list* has the identical information as an assembly list, however it has been sorted not by sequence number (how they edit together), but rather by ascending film numbers. By organizing the shots in this way, an assistant or negative cutter can pull shots in the order they appear on the unrolling film. This facilitates the breakdown of film for future assembly:

Courtesy of LightWorks

Edit:	SC 50 VER 2		Project: MRS. DOUBTFIRE		
Edit dated:	21:08, 15 Sep 1993		Format: 35mm		
Report dated:	21:31, 16 Sep 1993		Cutting-copy version 4 (6.1)		
Pull list by roll					

***NOTE: 1 case of material re-use in this list
Full report after pull-list

Evt	Roll	Shot name		Start code	End	Length
1		50-4		KJ458822 5117+10	5131+07	13.14
3		50-4		KJ458822 5136+07	5150+01	13.11
8		50-4		KJ458822 5173+10	5182+05	8.12
4		50A-2		KJ458822 5250+06	5261+07	11.02
6		50A-2		KJ458822 5262+11	5268+13	6.03
2		50A-2		KJ458822 5273+03	5275+12	2.09
5		50B-1		KJ458822 5301+02	5305+10	1.10
7		50C-1		KJ458822 5353+04	5357+00	3.13
10		50C-1		**KJ458822 5357+01	5357+02	0.01
9		BLK-	Cam	KJ458822 5411+02	5418+09	4.08

End of pull list

It is important to note that for pull lists and assembly lists, there will usually be two options: negative or workprint. The lists are identical except that the negative list is built from *key numbers* and a workprint list is built from *code numbers*. Consequently, workprint lists have a further option of printing *picture* numbers or *sound* numbers.

Another factor in film list printing is the *cut margin*. Although film editors splice workprint with 2 to 4 sprockets of clear tape, negative cutters actually melt adjacent frames together with a "hot splicer." For 35mm film, editors leave a single frame clear at all splices to allow the negative cutter room to make the splice. Even though this hot splice only destroys a sprocket or so of film, allowing for a full frame is considered safer. This boundary frame, called the *cut margin*, must be accounted for at all splices. Prior to list printing, editing computers require a value for the designated cut margin. When a computer checks an edited sequence for use of duplicate or adjacent frames, the cut margin must be factored in to insure the negative cutting will go smoothly. **Duplicate frame** use will be part of a pull list or be in its own separate *duplicate printing* list.

➤ An *optical list,* or optical count sheet, is a way to recreate on film those effects created and previewed in video. Many systems will locate effects in an edited sequence and generate a count sheet for an optical house to build the identical effect using film elements. These lists define the type of effect, indicate start and stop key numbers, and durations of the given elements.

```
Edit:          SC 50 VER 2              Project: MRS. DOUBTFIRE
Edit dated:    21:08, 15 Sep 1993      Format:  35mm
Report dated: 21:31, 16 Sep 1993       Cutting-copy version 4 (6.1)
Opticals list: printing handles 32 frames
```

	Start printing	Start out	Length	Full in		Stop
4	50A-2 (roll A19)		Dissolve	50B-1 (roll A19)		
	KJ458822 5248+06	5258+08	3.00	KJ458822 5304+00		5307+10
8	50-4 (roll A19)		Dissolve	BLK-	Cam (roll A19)	
	KJ458822 5171+10	5179+06	3.00	KJ458822 5414+01		5420+09

End of opticals list

Courtesy of LightWorks

➤ A *change list* is the least commonly found lists in general purpose nonlinear systems, but essential for electronic editing systems cutting film projects finishing on workprint. In the normal course of events, scenes are edited offline, electronically. Once scenes are cut, they will often need to be screened in a theater. This will involve the use of pull and/or assembly lists. However, once the cut film exists, changes made to the sequence on the electronic system will be somewhat difficult to re-create in the now-edited workprint. Because of this, a separate kind of list needs to be generated: one that will compare the original cut (which exists on workprint) to the current cut (created on the nonlinear system), and inform an assistant how to modify the workprint to bring it up to date.

Title: Sc1DC	EditDroid Picture Change List	Fri Jan 11 16:11:56 1991		Page 1
New Title: Sc1DC	Footage: 81+09	Duration: 54:11		
Old Title: Sc1	Footage: 77+01	Duration: 54:11		
Description: Director wanted changes.		Project: StarWalk		

Edit No.	Old Sync Pt.	Change +/-	New Sync Pt.	Type of Change Lab roll	Scene	Take
5	28+07	- 0+00	28+07	Trim Tail		
	DD63 7438+01		DD63 7438+01	14	21	3
	Boar approaches, Capt in FG.					
8	42+06	+ 2+12	45+02	Add Duplicate Shot - - Get From Reprint		
	B2X5 1141+05		B2X5 1144+00	2	18A	1
	Buffo in the field tree, ZOOM to CU.					

Many film projects only use electronic systems for first cuts (once they conform workprint, they do all additional changes on the film workprint), due to the general difficulty of revising workprint among other considerations; consequently, systems that do not have the option of a change list (also called a "diff" list as it highlights the *differences* between two film versions) can be utilized on those projects.

A change list will generally present the least number of modifications it would take to convert one edited sequence into a new one. The user will have to select the two lists — presumably an old and a newer version of the same sequence, and then "request" the system generate the change list. A change list is based on the fundamental property that when changing the original workprint into a new version, only two actions ever happen: any piece of film can either be CUT out or a piece can be INSERTED in. Exactly how these operations are peformed is critical. They can be done in two passes through the workprint (first to do all the cuts, then to do all the inserts);

or in a single pass (cutting and inserting as you go). Neither method is "better" than the other, however meticulous attention must be maintained to detail throughout the entire process in either case.

➤ The next category of computer printout might be called *system lists*, as they are hard copies of specific screens viewable on the editing system. These might include printouts of dailies listings, logging screens, and so on. There is a wide variety of these lists.

➤ The final category is a printout of *storyboards*. Many systems have the ability to laserprint cuts or dailies in storyboarded formats. These printouts are particularly helpful when working with clients familiar with storyboards — as on feature films or commercial projects. Although the image quality of the frames is not always high, it is sufficient for most productions to have a visual hard copy of material on the system.

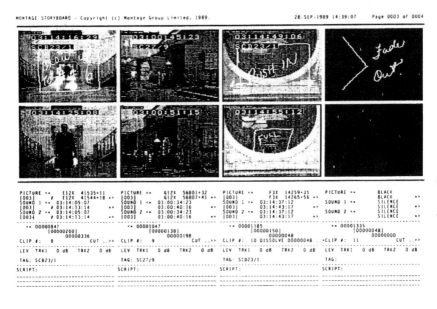

CHAPTER 5

THE SYSTEMS

T Y P E S   O F   E D I T I N G   S Y S T E M S

IN THE BEGINNING, there was film. Film editing, which we will call **TYPE I** editing, continues today using uprights and flatbeds. At first, even the dawn of videotape did not stop traditional film editors from cutting tape like film. For almost a decade, videotape was almost like another *kind* of film. **TYPE II** editing was characterized by computer-controlled linear systems and the birth of timecode; the media was usually videotape, but occasionally other things, like videodiscs. There was no database; there was one source reel for each set of dailies. There were the rudaments of nonlinearity in the form of a "preview" of the upcoming edit. **TYPE III** editing systems began the *Age of Nonlinear*. Multiple source tapes or discs were used to create random access and longer previews (extended nonlinearity). Databases. Graphic displays. **TYPE IV** editing was only a subtle, but important, variation on TYPE III; it produced the first virtual masters; extended previews, but with full-user control. It was only viable using laserdiscs. Computer-wise, it was very much like TYPE III — there was considerable scheduling of many source decks. But to the editor, it was a dramatic departure. **TYPE V** systems were radically different than previous ones — source material was now digital. There was no scheduling of decks, no previews, no shuttling. Although perceptually it looked like the TYPE IV, technologically it was a new entity. Like TYPE IV, it was sometimes called *virtual editing*. And like TYPE IV, and even to a greater degree, it was finally completely nonlinear.

People who follow developments in the Age of Nonlinear often refer to three generations of equipment evolution. Many followers of the nonlinear scene describe systems using this terminology; however, there are sometimes slight variations in the criteria for each.

First Generation

Before anything — before personal computers, before videotape, before linear editing — there was the CMX 600. It was so far ahead of its time in all ways that although it is the granddaddy of everything, not many people have ever seen or used one. In spite of its clear position as the first electronic editing system, linear or nonlinear, the CMX 600 is not considered part of the "first generation."

The first generation was the "new phase" of nonlinear systems that debuted circa 1984-85: the Montage Picture Processor, the EditDroid, and the Ediflex. Each system (and its respective manufacturer) had to contend with a world not yet sure of electronic nonlinear. Film editors were being thrown onto these systems, and no one had a good idea of the best ways to use them, or for exactly what market the products were most suited. These things would come in time, through improvements to the systems and the benefits of hindsight. These systems all embodied *type III* editing.

SECOND GENERATION

The second generation came on the heels of the first — a little smaller, a little cheaper, and sometimes designed with the perceived mistakes of the first generation in mind. The systems were the CMX 6000, the E-PIX, the TouchVision, the Link. The second generation was introduced to a community that was aware of the advances that had been made in editing. Episodic television, some films, movies-for-television, and commercials had all been tried with a great deal of success on the first generation. The new systems capitalized on the somewhat unsteady financial nature of some of the earlier manufacturers. The first generation continued to respond to the debut of the second with continued development and price reductions. These system continued to embody *type III* editing, with the exception of the CMX 6000 — the original *type IV* system. These systems were developed between 1986 and 1989.

THIRD GENERATION

The third generation of systems is based on personal computers and it is software intensive, but its distinguishing characteristic is that source material is *DIGITIZED* for editing. Third generation digital systems began with the EMC2 and the Avid. Instead of recording numerous VHS or Beta tapes or utilizing laserdiscs, source tapes are digitized into the system, where editing can remain in the digital domain. The Avid was introduced at roughly the same price as the second generation systems. The EMC, on the other hand, was the first system to approach the ideal for offline, *i.e.* low-cost nonlinear editing. Because for the first time there was never any master videotape, digital systems are sometimes called *virtual editing systems*. These *type V* systems debuted in 1989, and continue evolving and proliferating today.

The trend is towards market specialization and the newer desktop (low-cost) emergence initiated with the EMC2. Later, through systems from Adobe, Data Translations, and D-Vision, among others, professional nonlinear is approaching the ten to fifty thousand dollar range. Personal computers running software packages and requiring minimal hardware can provide independent film makers, commercial and industrial producers, and schools an avenue into nonlinear editing. From there, viable hybrids of low-cost linear and nonlinear systems can be used on many types of productions.

CRITICISMS OF DIGITAL SYSTEMS

The principal criticism of the third generation of nonlinear systems has been the picture quality and frame size of the images. Because of the technology and economics of digital video, there is a direct trade-off between image quality and source storage quantity. Since the quantity of storage you have available is largely an issue of COST, the question arises *"How good does an image have to be for an editor to use it? And at what cost?"*

In terms of computer data, digitized video is huge. The difference between broadcast-quality picture and broadcast-quality sound is about 100 times. But there are tricks to make the video take up less space: Make the pictures smaller. Decrease the resolution. Play the pictures slower. Develop digital compression techniques to take big pictures and make them take up less space.

All these methods have been used.

But still the question remains, and may never be answered definitively for the editing community: "How good is good enough?" Filmmakers often feel broadcast video resolution is too poor for them. Videotape editors cringe when looking at 4th or 5th generation dubs. Everyone likes seeing pictures large, on beautiful monitors. And yet, editors', directors', and producers' standards (although professed to be adamant) have been flexible. Directors and Networks watch dailies on VHS cassettes where they used to demand screening rooms and workprint. And in 1990 editors began using small, low-resolution digitized images for editing.

No one argues that digital images won't get better. But regardless of how good the images get, there will always be a trade-off between quality and storage.

Lesser criticisms concern the difficulty in simultaneously playing more than one video image. Digital systems have enough difficulty moving one image at 30fps — to move two or more images, interlocked (as you might in transition editing or multi-camera modes) is even more difficult.

Regardless, all criticisms are sure to slowly fall by the wayside as these technologies mature over the next few years.

DESKTOP VIDEO

The 3rd generation nonlinear systems combined with the ongoing decrease in costs of powerful personal computers and high-quality consumer camcorders (with tape formats like S-VHS and Hi-8) has spawned the growing world of *Desktop Video*.

Early on, the publishing world was introduced to personal computers. It had formerly been a mysterious land of typing on typewriters, authors' submissions, typesetting equipment, manual layout with waxers and X-acto knives. The new computers were able to handle an array of tasks formerly limited to slow, old-fashioned ways, and on occasion, expensive equipment. The 80s saw the advent of *Desktop Publishing* — where personal computers loaded with appropriate software could handle the creation of pamphlets, magazines, even books: from writing, typesetting, graphics and layout, to color separation and printing. By the end of the decade, the established publishing industry also began to adopt the methods and products.

The video post-production world seemed considerably less vulnerable: the technology of broadcast video appeared to be much more technologically difficult, considerably more expensive, and pretty much unknown outside of a few individuals with extensive knowledge.

For better and worse, those days seem to be drawing to a close.

Since it was first introduced in 1984, the power of Apple's Macintosh computer line has increased dramatically. Today, a Macintosh or Windows NT/95 computer with appropriate hardware and software can produce an astonishing display of video functions. Although computer video is unlike NTSC video — with the correct video converter card, the computer's video signal can be converted to NTSC and recorded onto traditional videotape. Also, image quality continues to improve; it's already at acceptable levels for delivery of many corporate and training projects. Since 1994, some television is actually *broadcast directly* from digital nonlinear equipment.

The available software and hardware range in price from a few hundred dollars to many thousand, and as quality and functionality increase, the systems are becoming less and less distinguishable from "professional" products. The most noteworthy systems are stand-alone "studios in a box," which either control videotape decks, and combine the video with effects prior to output, or are digitizing workstations that take in video, edit and process it, and output the "online" product.

The present machinery of video post, with nonstandard interfaces and complex design, require a highly trained "priesthood" to translate the ideas of directors and producers into video. Often the creative person is forced to speak *through* the editor, designer, or composer. But as the technology of video post-production becomes more simplified and accessible to creative individuals, the premium in the business will no longer be placed on technical skill, but simply on creative vision. The talent will be solely in a person's ability to organize ideas in picture and sound — a rare skill made that much more obvious by the transparent technology. The divisions of labor so familiar in the historical world of production and post-production may begin to wear thin.

In all their incarnations, the nonlinear electronic editing systems for the past decade have begun to do exactly this. They have upset the *status quo*; they have blurred the lines. It has proven to be a frightening prospect both to established editors, who fear for their jobs, and to facility personnel, who see the dramatic changes in equipment looming on the horizon.

QuickTime™

 In May 1991, Apple Computer Company introduced a new format for playing moving digital video on their Macintosh computers. For those familiar with Macs, it is a file format (as are PICT, TIFF and EPS). It is not a product. It is not an application. It is just a *method* by which digital video and audio can be moved, placed, and processed.

Following the announcement of QuickTime, many software developers began allowing for the QuickTime "movies" to be placed in their applications — moving pictures can now be pasted into spreadsheets, scripts, flow charts, music charts, mail, and so on.

QuickTime was designed to play full-screen (640x480) images, full 24-bit color, at 30fps — early visual limitations of QuickTime (160x120 pixel frame size) were really just limitations of the digitizing boards and the CPU power in many Macs. Newer hardware products like the Radius' Video Vision Studio, when combined with powerful Macintosh computers (like the Power Mac series) and appropriately fast hard disks or disk arrays, can capture and play QuickTime images at very high quality. These new high-end digitizing boards begin to move QuickTime video in the professional environment — and QuickTime is finding its way as a tool for previsualization of professional work, and even delivery of some corporate and commerical video.

There are presently QuickTime "film" festivals. Due to the affordability of desktop editing products, QuickTime is likely to be many future filmmakers introduction to video and editing.

PURCHASING SYSTEMS

Setting out to own a nonlinear system is often a harrowing experience. Since current rental prices for fully configured professional systems are in the $1,800 - $2,500 per week range, it often seems lucrative to own equipment outright rather than rent. Rental prices for high technology, for those untrained in standard economic paradigms, are loosely determined based on a weekly cost of 2% the purchase price of all the rented equipment — remember that the owner of a high tech piece of equipment wants it paid for in about a year.

Desktop systems are a fascinating and dramatic change to the professional industry. But fair comparisons are critical to come by and difficult to establish. A few key points to remember about systems and technology:

▼ All systems run on some kind of computer. A computer is basically a CPU and some memory. A Macintosh is a computer. An IBM XT is a computer. A Silicon Graphics Indigo is a computer. And there are many many more you may not have heard of. Editing itself, it has been established over the past few decades, does not require a particularly fast or powerful computer. Digital video, on the other hand — playing it, augmenting it, manipulating it — requires considerable power and speed. Even some computer interfaces (like bit-mapped graphical displays) require more computer power and sophistication than boring character displays. What's the point? Whether you buy a pre-packaged system or you buy readily available components and put together your own, you generally need the same stuff to perform the same functions.

Today, top-of-the-line personal computers that have enough processing power and size to run editing applications are top of the line personal computers — they might cost from $3,000 to $5,000 dollars.

▼ If the system digitizes video it needs hardware ("boards") that digitize. Lots of companies make them. A few are more commonly recognized because consumers buy them, but others are known more to professionals. A good quality digitizing board that does some digital compression might cost from $2,000 to $6,000 dollars.

▼ If the system reads and recognizes SMPTE timecode, it needs a board that is a timecode reader. Even professional videotape machines need special circuit boards to read (timecode readers) or record (timecode generators) timecode. And there are a few kinds of timecode (VITC, LTC, etc.), each with different performances and associated costs. Bottom line: standard timecode readers cost between $500 and $2,000 dollars.

▼ Videotape. Input and output. Some kind of video must be recorded into the system. Therefore, you need either a videotape deck or camcorder as a source and/or record. Professional decks for offline are usually 3/4" VCRs or Beta SP. Prosumers tend towards the growing quality of Hi-8 and S-VHS. (Professional 3/4" decks start around $8,000. Hi-8 and S-VHS cameras are about $1,000 and up and decks start around $500.) Machine control on input or output requires a professional deck with timecode.

▼ Machine control is required for most kinds of video input digitizings or videotape output recordings. These can be specially built or off-the-shelf controllers (*e.g.* utilizing V-LAN protocols), and there are many degrees of sophistication here (to none at all, as many system have no machine control). A decent machine controller costs about $1,000 dollars. Some machine-control devices also read timecode.

▼ Monitors. Almost all digital nonlinear systems output computer video — but not necessarily NTSC video — so you can't always just hook up the old TV set to your new editing system. There are lots of kinds of monitors. TV monitors (affectionately known as NTSC monitors) do not display computer video. Multi-sync monitors (somewhat complex, expensive and versatile) can display computer video, data and NTSC video. Bigger monitors don't necessarily mean better looking pictures. A digital picture can be tiny or giant with no change in resolution — it will just look terrible blown up. Monitors can accept VGA or RGB, but would need special cables, adapters or sometimes converters to play any kind of video it wasn't designed for. The cost of mid-sized professional video monitors (17 inch is on the small but acceptable side; 21 inch is very nice) is anywhere from $400 to $3,000. Multi-sync monitors are slightly more expensive than non-multi-sync monitors of the same size. For whatever system you use, you will need at least one computer monitor, sometimes two, and often a video monitor (which is also good for watching the output of your video deck).

▼ Storage. Storage costs money. More storage, more money. Faster storage, more money. More compact storage, more money. Hard disk storage, a common choice for a variety of performance and economic reasons, has been growing in capacity recently, and a little more economical. Fixed hard disks (consumer and professional alike) cost less than $400/ gigabyte and even that is dropping fast. But simply being big enough is not sufficient for any storage medium. It must also have a minimum set of performance parameters to do the job. Among them, sustained data transfer rates of more than 3MB/second. Lesser performance will generally translate into lesser quality video.

Another thing to remember when you purchase hard disks is that these disks must be *formatted* prior to use. You might imagine a new hard disk like a freshly paved parking lot: no lanes. To make the lot efficient and ready for use, slots must be painted in. Formatting records some critical data onto the drive to prepare it for use — the bigger the disk, the more space needed for the required formatting. For example, 2MB floppy disks format down to hold about 1.4MB; 2.4GB drives format down to hold about 2.0GB. When you get a drive, what's important is what it holds *after* formatting.

Desktop systems often appear significantly less expensive than their professional counterparts. Remember to assess cost for systems compared with identical hard disk storage on-line. Common desktop configurations utilize a single large external hard disk — for example, the Micropolis 2.4GB drive — which provides about 40 minutes of video and audio, full screen, 30fps, at a reasonable but not exceptional offline image quality. High-end systems are often high end only because they have 10 or more hard disk drives accessed simultaneously.

LOW-CALORIE HARD DISKS

Let's say you're in a grocery store, shopping for diet foods. You pick up an 8 oz. fruit drink, and look at the label. Its says that there are 12 mg of sodium and 4 mg of fat per serving. Sound good? *It depends on what a serving is.* Most people might not realize it, but a serving is not a bottle — it's a few sips. It says in fine print that for this drink, 3 oz. is a serving. So they kind of mislead you, without actually lying, about how good this drink is for you. Get it? If something is too fattening, to make it look better, they just make the "serving size" smaller. For digital storage, it is exactly the same thing. To make professional editing systems sound better, manufacturers might announce they sell a "4-hour" system or a "10-hour" box. But they could simply be making the image resolution slightly less good (thus holding more) to make the claim. Comparing two "30-hour" editing systems is not appropriate unless you have fixed both at the same image and audio quality.

Summary: To configure a minimum quasi-professional digital system, you'll need to round up the following:
COMPUTER
MONITOR(S)
VIDEOTAPE DECK (Input/Output)
MACHINE CONTROL/TIMECODE READER
DIGITIZING HARDWARE
HARD DISK (or other) STORAGE

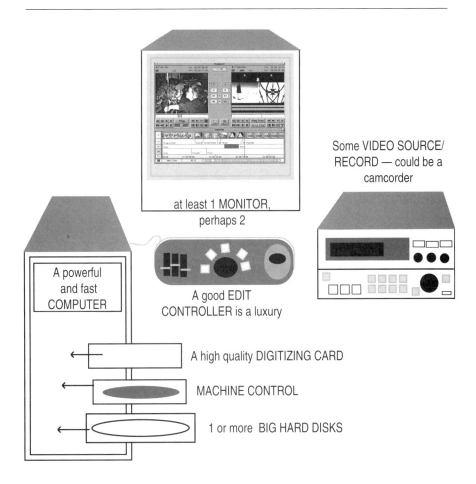

Some VIDEO SOURCE/
RECORD — could be a
camcorder

at least 1 MONITOR,
perhaps 2

A powerful
and fast
COMPUTER

A good EDIT
CONTROLLER is a luxury

A high quality DIGITIZING CARD

MACHINE CONTROL

1 or more BIG HARD DISKS

Not counting cables and furniture and various and sundry peripherals, you could easily spend $15,000 for a minimal system, and $30,000 for a more decked out one. The difference between this home-configured system with off-the-shelf software (from companies like Avid, Adobe, Data Translations, D/Vision and others) and professional high-end systems (which generally offer some combination of off-the-shelf and proprietary equipment) is often simply a matter of reliability, service, and interface style. This last issue, *interface*, is perhaps the most important, as it will determine the speed, ease of use, and overall power — issues central to the professional, and more ancillary to the pro-sumer. Professional systems tend to have dedicated editing controllers, specialized tools for producing a variety of input and output options, and numerous ways to manage large volumes of source material, often frustrating on desktop systems.

SYSTEM HISTORY OVERVIEW

This section will serve as a specific review of products, placed in historical and technical perspective, in roughly chronological order. Unlike the more corporate discussions in Chapter 2 ("A Brief History of..."), this review will not be concerned primarily with economic success or failure, or realistic design strategy — only with fundamental form and interface.

Let us begin by looking at computers and a part of their evolution that involves us. In the 1970s and early 1980s, work was done at Xerox's computer research division in Palo Alto, called Xerox Parc. There, a number of innovative computer interface technologies and devices were pioneered, in particular: the bit-mapped display, windows, the icon, and the mouse.

In 1981, where most computer companies or designers in the world were working from more standard PC-type monitors and technology, a few saw a different way for computers to interface. The first three nonlinear systems in our discussion are all extremely unique and innovative: (Lucasfilm's) EditDroid, (Cinedco's) Ediflex, (the Montage Group's) Montage Picture Processor.

Although the early Ediflex prototype had been in some use prior to the release of the other two, all debuted as of NAB 1984, and tested on their first projects during the following year.

photo courtesy of LucasArts Entertainment Company

The EditDroid (1984) shown with the "Death Star" console, as it was affectionately known. Note the central rear-projection widescreen display for the edited master, with the source monitor on the left and the SUN data monitor on the right.

The EditDroid "TouchPad" — a stand-alone console for editing. On the right is a trackball and select button, on the left are "soft-function" keys that change as the editor performs different functions. In the middle is the metal "KEM-like" shuttle knob, and four buttons for editing. Editing took place largly on the data monitor via the trackball.

The first EditDroid worked primarily with 6 laserdisc players and two 3/4" decks. It got most of its random access from the discs, but could also make use of videotape for laying off parts of the cut, and for accessing source material not yet placed on laserdisc. The EditDroid ran on a SUN workstation, the highest powered platform of the available systems, and a bold step for 1984. (The SUN platform had actually be selected in 1981.) The system had a high-resolution bit-mapped display, a graphical timeline that was used in editing (and not simply in representation), a highly designed editor's "TouchPad" that included a KEM-style knob for shuttle, a trackball, and some variable function buttons labeled with LEDs that changed as you worked. The cursor, as you moved across the screen, was a little book in the "Log Window," a scissors in the "Edit Window," and a tiny Darth Vader mask in the thin "Control Window" along the bottom of the screen.

Even with the laserdisc source, material needed to be scheduled and previewed — it was still not possible to roll material in a preview without recording. Effects were performed via a Grass Valley 100 switcher that was configured with the system.

The EditDroid computer display — the top portion of the screen is the LOG WINDOW, and below that is the EDIT WINDOW. The Log Window contains a listing of shots on any given laserdisc; the Edit Window contains the "cutting block" where picture and 2 channels of audio are cut, and a timeline (running right to left) where shots are placed. Below the timeline is a text display corresponding to the timeline, and along the bottom of each window are specific function buttons.

The Montage I controlled 17 Beta 1/2" hi-fi decks to achieve its random access. Material was broken into "loads" of up to 4 hours of source, and the metaphor was like hanging film strips in a film bin. In a complete departure from videotape editing, there was no "source" and no "master," *per se*; only big long strips that could be easily clipped apart into smaller strips, each hanging in the right order in a virtually infinite bin.

Thus the physical interface displayed 7 of the shots in a bin at one time — the head of the strip on top and the tail of the strip on the bottom. This "head/tail" view proved exceptionally useful in identifying long pieces of film. The editor's console was designed like a stand-alone machine, echoing the vibe of the film editor's Moviola. It was tethered to the racks of beta machines, but it sat alone, upright, with a futuristic console face that included two large, eminently tactile and responsive black knobs, used for a number of functions, among them for rolling your view of the hanging shots. Probably no system since has created such a tactile physical space for editing.

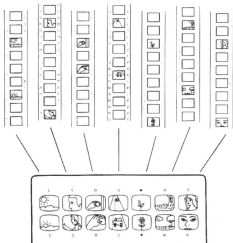

An illustration from the Montage User's Manual showing the fundamental concept of Head-Tail views in a bin. The bottom part shows the 7 monitors from the console.

Below is the original Montage console, with 3.7" b&w monitors for the bins (in the middle section), the NTSC edited sequence (top right) and three 5" monitors for selected trims and splices, and a status display.

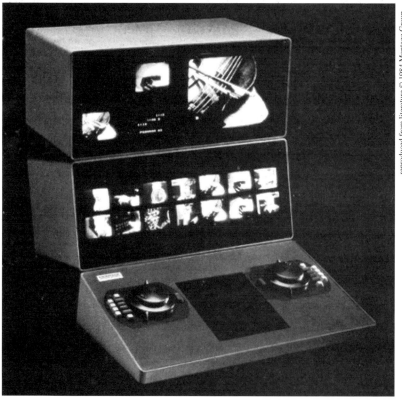

The racks of 17 Beta decks used with the Montage.

Also quite unique was the Montage's use of digital pictures: rather than wait for the Beta tapes to cue to any given head or tail, the Montage digitized the head and tail frames "labels" and stored them on a separate 10MB Bernoulli hard disk that went along with a project. When a shot was identified via the digital picture label and selected, the system then searched the tape for the identical frame and once cued, replaced the digital image with the videotape's color frame. This way, labels existed to represent all shots in all bins, even without relying on the videotape's ability to cue.

The cue time to play a preview ranged from a few seconds to a number of minutes, in some cases.

Unlike other systems of the age, the Montage also allowed the editor to scratch notes on the digital picture labels with an "electronic grease pencil" — a pen and digitizing pad built into the console. All material input into the Montage is stored in the "source bin" where no editing can take place. From there, material is discarded from (to the "discard bin") or copied to. The editor builds cuts into any of four separate "work bins" or in a "pull bin."

The Ediflex was a third and still different kind of system. Tailored in a way for episodic television editors, built by the only person with professional editing system design experience (*Adrian Ettlinger*) and the only system from Hollywood. Where the EditDroid represented source material in a powerful configurable database (a list, in other words), and the Montage used its unique head/tail labels to represent everything, the Ediflex was based on the script. A lined script is the fundamental basis for source information — created mostly on the set during production, it is essential to most script-based editing — and a natural for the computer extension of source material. Although the process of getting the script into the system and correlating it to the source footage was a bit arduous for assistants, it

gave editors unusual functionalities. An editor could easily call up a given line of dialog from every take that covered the line. This was called the "Script Mimic".

All of these systems, so completely unique as you might expect from three inventions made simultaneously with such different goals and hardware, were attempting to attract film editors to video, even though those editors had rejected the move during the previous decade. They were trying to re-create the film **style** with better computers and less expensive video technology. All were expensive for the time, with fully configured systems running from over $100,000 to $200,000. The source media for the Ediflex, like the Montage, was 1/2" videotape, and of inconsequential cost. The EditDroid required the use of laserdiscs, both expensive (compared to tape) and difficult to come by. The device used to record these write-once 30-minute laserdiscs was the Optical Disc Corporation (ODC) RLV recorder, itself costing more than $175,000.

The EditDroid, although arguably technologically superior to its competition, was labored with huge costs and barriers to viability (in the form of disc-dependence). The Montage began making inroads on long-form — television and movies — in various locations around the world (sales to big-name directors — Coppola and Kubrick — gave the Montage its biggest boost); the Ediflex settled into the Hollywood market as a rental-only system.

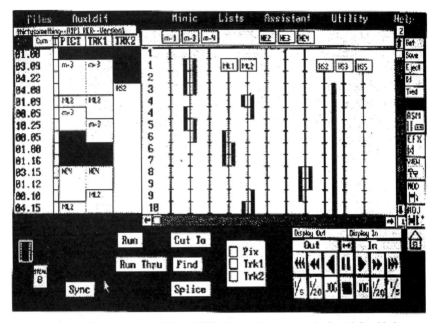

The updated interface of the Ediflex (1989): the timeline runs vertically at left, with the "script mimic" along the right. Takes run vertically through numbers representing each line of dialog.

Although the interfaces were quite different, The EditDroid, the Ediflex and the Montage all scheduled previews, in much the way traditional videotape editors had previews: you chose a sequence or portion of a sequence, the machine would schedule the shots (either from the tape sources or the very fast laserdiscs) and then show you the preview. The editor had no control of the preview — it was shown in play-speed forward. The EditDroid enjoyed the luxury of quick laserdisc cueing, often a fraction of the time required by the tape-based systems.

In 1986-87, CMX released the CMX 6000, a laserdisc-based system that was considerably different from its predecessors. Although the simple style and dedicated console are often cited as the central features of the system, it was the *virtual master* that truly advanced the state of nonlinear editing. Laserdiscs cued forward or reverse as you shuttled the "master" around, which simulated an actual edited sequence.

The system had moved very far in removing the computer from between the editor and the material; where the EditDroid was heavily computer dependent with its windows and icons, the Ediflex to a lesser degree, and the Montage the least, the CMX 6000 was very list-oriented, but removed a great many computer functions (both to its credit and to a fault).

The greatest drawback of this system was ironically its defining characteristic — its solitary use of laserdiscs. The discs gave the system the speed and interaction no prior system had, yet forced the user to be dependent on ODC disc pre-mastering. This process was both time consuming and expensive, and discs, once written, could not be erased or added to.

Still, the system was based on the very simple idea of locating source material in one monitor, and attaching it via a "splice" button to the master in the other monitor. Overlaps and inserts were particularly fast and easy.

Dependence on the computer interface is really not much of an issue

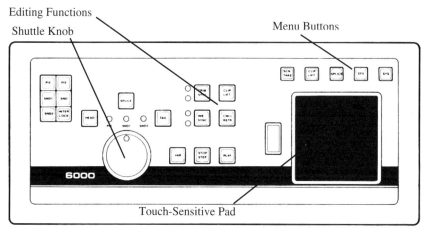

The 6000's Edit Controller

Master (left) and Source (right) Monitors; Master monitor doubles as the outgoing frame in transitional editing — the Source monitor becomes the incoming frame. Source also serves as a computer interface display when not running dailies.

photo courtesy of the CMX Corporation.
© CMX Corp.

until you have a bit-mapped or icon-oriented display — the EditDroid was years ahead in this department, not seen again until the digital systems.

The BHP TouchVision had a technical architecture much like the tape-based Montage and Ediflex. But its interface again was unique. Instead of hiding the multiple sources that the computer relied on for random access,

TouchVision system.

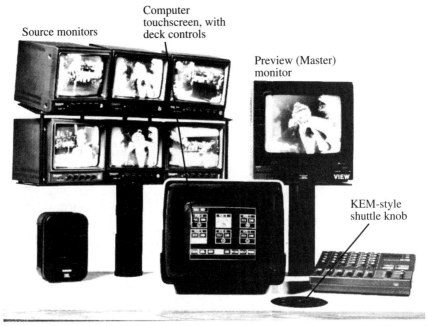

Source monitors

Computer touchscreen, with deck controls

Preview (Master) monitor

KEM-style shuttle knob

the TouchVision changed that perspective to a film-reel metaphor. Now, reels could be loaded onto individual decks, and controlled in much the same style as a multi-plate flatbed editing table. The system console was a dedicated tabletop with a KEM-style knob built in, further re-enforcing the filmic flavor. Although any of the tape-based systems could manage a multi-camera video or film editing style by dedicating a certain tape to a certain camera roll, multi-camera required a number of specific functions that would leave many systems un-utilized for these projects. Eventually, with later releases of software, both the EditDroid and the CMX 6000 could also manage multi-camera projects, but due to disc economics, they remained most viable for single camera (and ultimately) film rather than video source.

One system that was released with the original nonlinear products was an in-house system called the Spectra-Ace. Familiarly known by the name of the facility, LaserEdit, the product was not nonlinear but only random access due to its use of laserdiscs for source instead of tape. Shots were accessed via timecode, manipulated much like videotape, and recorded to tape. The large, video-styled console generally did not facilitate the easy transition for film editors, so in time the system remained the domain of video-savvy editors, and became the predominant system in Hollywood for multi-camera television shows for many years.

Amtel's E-PIX sought to gain the advantage of the disc-based systems (fast access, high-quality images) while removing the dependency on the ODC laserdisc process. By creating a system with *recordable* videodiscs, this was solved, but only partially. The high cost of these WORM videodisc

E•Pix Edit Console

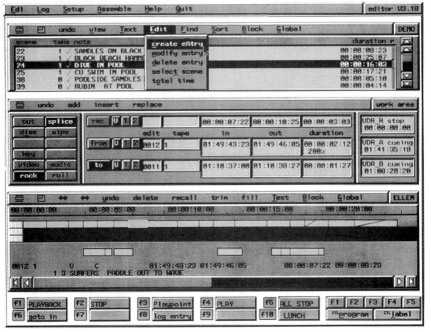

E-PIX's version 4 with graphical display (1991).

machines required that few be used on the system, which in turn required the evolution of clever new disc-use strategies. The basic premise was as follows: disc real estate is limited, expensive, and should not be squandered. Rather than transfer **all** source material to multiple copies of tape or disc, the E-PIX transferred selectively to the discs — in effect building "super select" discs of required shots. This concept was mirrored later, to a lesser degree, with the digital systems, as hard disk space is also somewhat limited and expensive. The E-PIX display had adapted a style not unlike the EditDroid — a cross between traditional videotape and a graphical interface. The timeline is a focus of the system, but so is timecode and timecode functionality. This makes the E-PIX video-like in its interface, like traditional linear systems or perhaps the Spectra System. Its console reflects this concept as it is neither particularly innovative nor simple. Its bit-mapped monitor is busy with functions and timeline — but it was desgined only to display information, and was not meant to be "active". The E-PIX was largely ignored in the small markets originally investigating nonlinear editing, but eventually found a serious niche: by the early 90s, professionals were not ready for the still-evolving functionality and poor image quality of the new

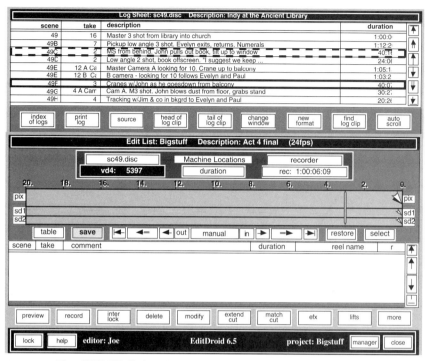

The editDROID display, after re-release version 6.5 (1990). Notice that the cutting block from the original system has been removed; System Window along the bottom has been expanded.

digital system but were growing tired of the old nonlinear technology. The videodisc-based E-PIX took advantage of the gap between the analog and the digital worlds, and like the laserdisc-based CMX 6000, flourished to some degree during the transition years.

By the late 80s, a few companies began investigating the possibility of *digital* offline. Both the EMC2 and the AVID began as projects from frustrated computer graphics individuals dealing with tedious and slow linear tape editing. Evolved from the corporate market, rather than from (or for) Hollywood, these two systems looked to personal computer platforms and off-the-shelf technology for much of their design; their primary goal was to make the editing accessible, to liberate it from the "high end." The EMC2 was positioned with a PC-based platform, common in American business; the Avid was initially developed on the Apollo graphics workstation but was ultimately introduced on the Apple Macintosh platform, less common for business, but growing in appeal with computer-friendly arti-

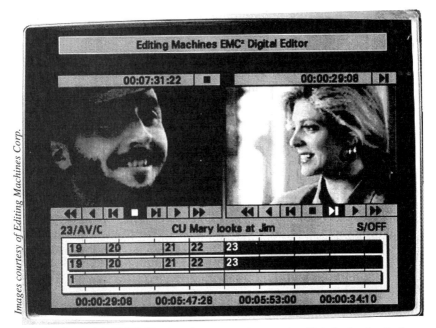

Early release of the EMC2. Note how the event numbers are placed into the timeline display.

Images courtesy of Editing Machines Corp.

sans, and leveraged with the massive religiosity of the Macintosh product followers. In a way, the chosen platforms set up the separation between the two products' thrusts: the PC for the low-end, and the Macintosh for the high-end — for the first many years of the two products, the EMC had been consistently less than half the cost of the comparable Avid line.

Although the existing manufacturers had long expected digital video to make its way into offline, they were confident their markets would not accept the image quality possible at the economics required. Avid and EMC bypassed the resistance mostly by avoiding Hollywood, and the markets the other systems worked in. The response was unprecedented, at least since videotape and computer controllers were first introduced for offline in the 70s. Where the sum total of Montages, EditDroids, CMX 6000s, E-PIXs, TouchVisions manufactured and used from 1985-1990 was perhaps under a few hundred, the numbers of EMCs and Avids sold in their first few years surpassed that, and by 1993, their numbers were an order of magnitude beyond all other nonlinear offline products.

The EMC, the first digital nonlinear system, was careful to rely mostly on removable media, magneto-optical disks, for source. Although this

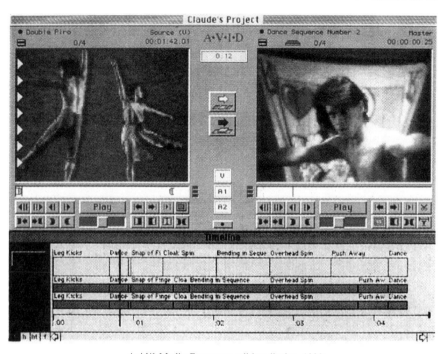

Avid/1 Media Composer, editing display, 1990

limited the image quality in particular, it was the most realistic option available for a flexible system. The interface was primarily a timeline, and something like a slide viewer representation of source. Both EMC and Avid licensed the Montage head/tail labels for their systems. The EMC invented editing primitives that were perhaps more complex than required for the functionality, but the system was known as efficient and cost effective. The Avid moved editing the farthest into the *computer* style (the Macintosh style, in particular), by windowing not only all the data and graphics (like the EditDroid, and later the E-PIX), but also all source and master material. The Avid had the most highly refined computer interface, well thought-out, and incredibly efficient for a software product. In spite of the company's efforts to offer various physical consoles for the Avid, it continued to appear that a mouse and keyboard were still the easiest and most streamlined ways to work with the system.

By riding the multiple platform options of the Macintosh line, and by bundling their software in different ways, Avid has been able to differentiate

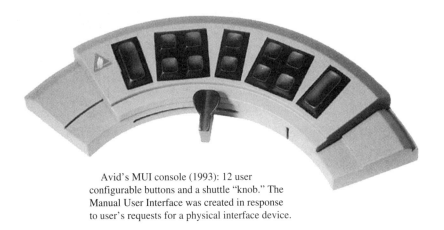

Avid's MUI console (1993): 12 user
configurable buttons and a shuttle "knob." The
Manual User Interface was created in response
to user's requests for a physical interface device.

a product line far beyond the scale of any other nonlinear company. Avid's marketing expertise went far beyond any previous nonlinear product manufacturer; bold advertisement headlines like "the first digital nonlinear system to play at 24fps," relied on the publics' still-naive views of nonlinear systems and positioned themselves as the originators of a type of product that was many years in development. Regardless, Avid succeeded where numerous others had failed — they proved that nonlinear video editing was both plausible and viable, and they introduced systems on a scale that was unprecedented.

In response to the immediate acceptance of the EMC and Avid, the earlier nonlinear system manufacturers began the quick reconstruction to digital. Three companies in particular took their established styles, their markets and expertise and invented all-new digital systems — windowed, bit-mapped, highly graphical. TouchVision produced the D/Vision; Montage produced the Montage III, and Ediflex produced the Ediflex Digital.

The D/Vision bore little resemblance to the TouchVision; it had a simple video metaphor that was easy to learn, and was reasonably functional. Competing with the lower cost systems like the EMC, the D/Vision was the first digital product truly replacing its analog product line. The Montage III removed the numerous video monitors used in showing the cut and the head/ tail views, and windowed them into a display. With the addition of a timeline, they combined two separate metaphors into the system: the original "film hanging in bins" logic, and the more-common horizontal timeline. The system added extensive database and organizational functions into the old Montage style, and used the familiar Montage console, but could

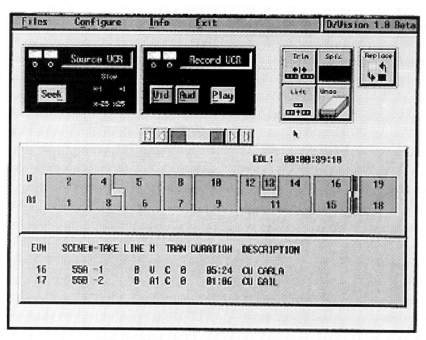

D/Vision editing monitor, system prototype version, 1991

be run with a mouse as well. It is unclear whether one method is more efficient than the other; certainly the mouse version was more economical. The Montage Group invented a new kind of technology to break the logjam between fixed hard disk storage — with low cost per hour, good access and image quality — and magneto-optical storage — removable, archival. Montage developed a RAID magneto-optical striping technology that used four MO disks in parallel, increasing the effective quality possible with usual MO drives, but maintaining their removability. Numerous technical issues plagued the initial release of the system; it was shown at many trade shows for several years before its first use in 1993.

The Ediflex Digital was a re-invented system, with many of the flavors of the Ediflex "classic", but with the elegant addition of Link Editing System functionality (the project leader was one of the inventors of the Link system). The Ediflex Digital, like the Link, was based on the shooting script metaphor for source material. The "mimic" process had been re-worked such that assisting is considerably less arduous than it had been, and the graphical display harkens back to the Link as much as it did to the Ediflex.

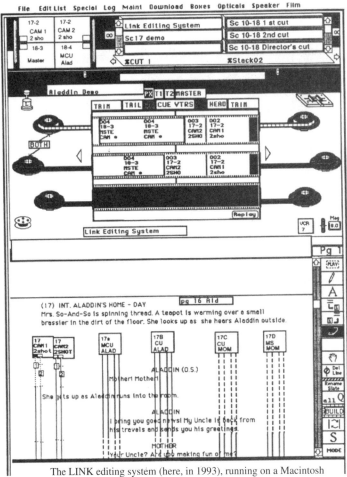

The LINK editing system (here, in 1993), running on a Macintosh computer, is still rented in Hollywood, even though only a few were ever built.

The vertical graphical timeline for the cut was unusual for picture editing systems (although more common in traditional audio workstations), but solved some of the "heads on left or heads on right" problems in timelined systems. Some of the graphical functionality associated with synchronization of tracks was quite unique, but the system had adopted a screen "toolbox" that required mouse manipulation for much of the editing functionality. The Ediflex Digital was prototyped early in 1993, but the project was canceled and the company closed in the following year.

Unlike the other digital re-inventions of the analog systems, CMX began modifying their CMX 6000 system to capitalize on digital source

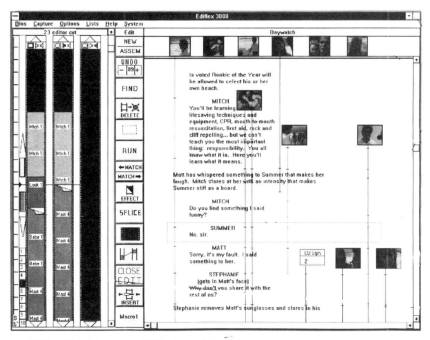

Ediflex Digital (prototype 1993). Areas of the display include the "Picture Gallery" (displaying the first frame of every take available at selected points in the script), the vertical timeline along the left has scene names inside each shot, and up to 8 tracks of audio all synced to the script), and "picons" (or slate boxes of scene/take info) used to represent each scene in the script mimic.

rather than laserdiscs, but otherwise maintained much of the hardware and software for the release of their re-named CMX Cinema. The Cinema then maintained the computer-low interface of the 6000 and the dedicated console while reducing hardware size, and adding, in effect, two giant digital "laserdiscs" — each capable of holding up to 26 hours with then-current hard disk technology. By adding specific tools for managing source material and a few much-needed functions to the architecture of the 6000, CMX augmented their analog system without actually inventing a new editing system. Of the re-designed analog systems, the Cinema was the only one to have evolved from its ancestor rather than to have been created new, in total. CMX terminated their digital nonlinear effort in 1994 and ceased development of the Cinema.

Once the window concept of editing was popularized by the Avid, other software companies began creating their own nonlinear systems — manipulating the various variables implicit in digital systems and applying cosmetic touches, both entirely driven by very specific markets.

D/FX was a company that had for years been producing high-end video equipment, in particular, their award-winning graphics workstation, the Composium. In a move to enter the nonlinear area, D/FX introduced a low-cost editing package called Video F/X which in many ways resembled an Avid interface. Their software-only option, Soft F/X was designed to open the way for post facilities to act more as "service bureaus," where you would get source digitized for you, and then you take it home to edit on your "satellite" system. The Video/FX was not fully released and quickly evolved into the *Hitchcock* system in 1993. Soon after, D/FX was closed and their Hitchcock system was sold to Aldus (pioneers of desktop publishing and inventors of such software as *PageMaker*), where a small team continued its development as a high-end desktop system. But when Aldus was purchased by Adobe (another desktop publishing pioneer, inventor of the Post Script printer language, and manufacturer of such popular software applications as

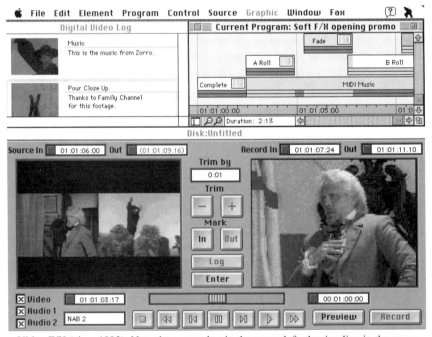

Video F/X (circa 1992). Note the source log in the upper left, the timeline in the upper right, and the windows for source (left) and master (right). The Video F/X was designed for the Macintosh platform.

Photoshop and Illustrator) in 1994, Adobe ultimately chose to stop development and disband the effort.

At the pure-software level, a number of Macintosh products have been introduced that can edit nonlinearly. The predominant entry here is from Adobe. Called Premiere, this software has grown quickly and remarkably since it was first introduced in 1992. Offered for under $1,000, Adobe's Premiere is based on a simple A-B video editor, but expands that idea to multiple video and audio tracks. Although it is not a particularly efficient editor, and is awkward at managing large volumes of source material, its strengths lie in its effects capabilities and its widespread usability for consumers and pro-sumers just learning about editing. It is less of an editing system and more of a desktop post-production studio — handling the integration of pictures and sounds and graphics impressively. Essential to this software package is the purchase of a high-powered Macintosh (like a Power PC or Quadra series), a big hard disk, and a digitizing board — image quality being solely a function of the quality of the board. Originally, the SuperMac Video Spigot made Premiere possible, but the release of the Digital Film card followed by Radius' VideoVision series of products made it viable. Premiere opened the door to "unbundled" editing software: allowing consumers to pick and choose appropriate CPUs, hard disks, digitizing boards, videotape decks, and so on. Prior to this point, nonlinear editing involved SYSTEMS, a combination of proprietary software and hardware. Following Adobe's lead, a number of manufacturers began offering unbundled software for the low-end users.

Adobe Premiere v.2 (1993): Examples of some of the window displays — the effects window, the construction window, and the source clip window.

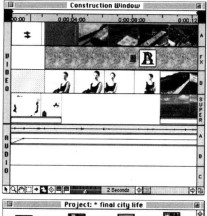

SYSTEM SURVEY

Due to the phenomenal popularity of digital nonlinear editing systems, and the vast proliferation of readily available magazines articles and product literature, it would be foolhardy to continue in this book's tradition of listing and editorializing on each available product's functionality. Editing systems simply change too often to be commited to the pages of a handbook. If you are seeking a convenient buyers guide or comparisons among the systems, check out the magazines listed in the appendix of this book (look for their most recent special issues dedicated to "nonlinear products").

Still, it is often nice to have a listing of products and manufacturers, and an oportunity to do a quick cosmetic inspection of their interface and overall layout, especially as this book considers interface paradigm greatly under emphasized as compared to the obvious issues of functionality.The next few pages are dedicated to this survey.

SPECIAL NOTES

This is by no means the total extent of systems that fall into the category of "nonlinear". Especially when you go beyond traditional definitions (horizontal vs. vertical), more and more products fit the bill. Similarly, nonlinear systems have evolved from dedicated hardware/software products to off-the-shelf hardware and unbundled software products; comparing them is a many splendered thing.

17 products are shown on the following pages; below is a listing of a few additional manufacturers that for technical reasons did not make it into the following survey, but that the conscientious shopper should include in basic research.

Broadcaster Elite	*VideoVision Telecast & Edit*	*Video Action Pro*
Applied Magic	Radius	Star Media Systems
1240 Activity Dr. Ste. D,	1710 Fortune Dr.,	1163 E Ogden Ave. Ste.
Vista, CA 92083	San Jose, CA 95131	705-364, Naperville, IL
(619) 599-2626	(408) 541-6100	60563 (708) 305-9432
VideoToaster Flier		
NewTek, Inc.	*Off-Line*	
1200 SW Executive Dr.,	Softimage	
Topeka, KS 66615	3510 St.Laurent Blvd.,	
(800) 847-6111	#500, Montreal, Quebec,	
	H2X 2V2 Canada	
	(514) 845-1636	

Manufacturer:	Adcom Electronics, Ltd.
Product:	Night Suite
Introduced:	E-Pix: NAB exhibition in 1988/Night Suite NAB 1994
Address:	310 Judson St., unit 6, Toronto,
	Ontario, M8Z 5T6 Canada
Phone:	(416) 251-2748 Fax: (416) 251-3977
Platform:	PC

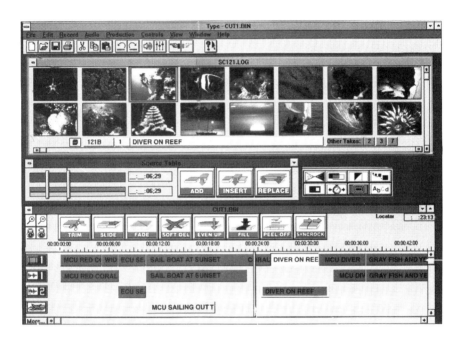

Manufacturer:	Adobe Systems
Product:	Premiere 4.0
Introduced:	1991
Address:	1585 Charleston Road, PO Box 7900,
	Mountain View, CA 94039-7900
Phone:	(415) 961-4400 *Fax:* (415) 962-2930
Platform:	Macintosh

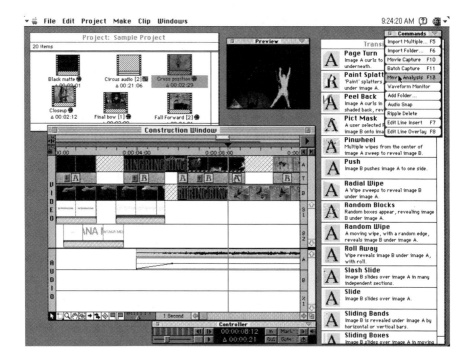

Manufacturer:	Avid Technology, Inc.
General Product Line:	Media Composer, Film Composer, NewsCutter, Media Suite Pro, Video Studio
Introduced:	NAB 1989 in Las Vegas
Address:	Metropolitan Technology Park, 1 Park West, Tewksbury, MA 01876
Phone:	(800) 949-AVID *Fax:* (508) 640-1366
Platform:	Macintosh, (MSP only: Mac, SGI, PC)

Manufacturer: Data Translation — Multimedia Group
Product: Media 100
Introduced: NAB 1993
Address: 100 Locke Drive, Marlboro, MA 01752-1192
Phone: (508) 460-1600   *Fax:* (508) 481-8627
Platform: Macintosh
(turnkey systems sold by Grass Valley Group, as *Video Desktop Personal Producer*)

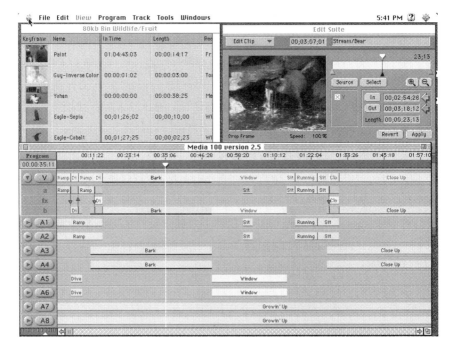

Media 100, v.2.5 (above); with detail of
editing window (right).

Manufacturer:	D-Vision Systems Inc. (formeraly TouchVision, Inc.)
Product:	D-Vision OnLINE, FilmCUT, PostSuite
Introduced:	TouchVision in 1985, D/Vision in 1991
Address:	8755 West Higgins Road, 2nd Floor, Chicago, IL 60631
Phone:	1 (800) 8-DVISION *Fax:* (312) 714-1405
Email:	71333.1761@compuserve.com
Platform:	PC

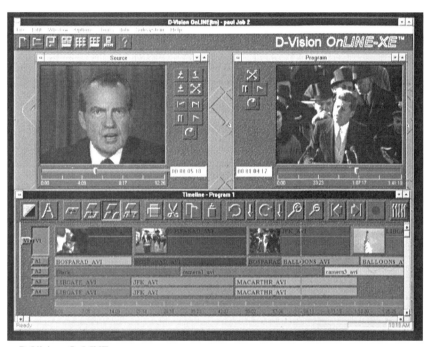

D-Vision, OnLINE

photos courtesy of D/Vision Systems, Inc.

Manufacturer:	Dynatech Group
Product:	EMC PrimeTime Editor, EditStar
Introduced:	EMC: SMPTE 1988
Address:	5125 MacArthur Blvd., NW, Suite 31,
	Washington DC 20016
	4750 Wiley Post Way, #150, Salt Lake City, UT 84116
Phone:	(202) 362-3102 *Fax:* (202) 362-3240
Platform:	PC

The EMC Primte Time editor (above); The EditStar, a specialized news cutting system that is text-based (right).

Manufacturer:	FAST Electronics U.S. Inc
Products:	Video Machine, Video Machine Lite
Introduced:	CeBIT 1992, Germany
Address:	393 Vintage Park Drive 140, Foster City CA 94404
Phone:	(800) 248-FAST *Fax:* (415) 802-0746
Platform:	PC

Manufacturer:	ImMIX, a Scitex Company
Product:	VideoCube, TurboCube
Introduced:	NAB 1993
Address:	PO Box 2980, Grass Valley, CA 95945
Phone:	(916) 272-9800 *Fax:* (916) 272-9801
Platform:	Macintosh (front-end, proprietary back-end)

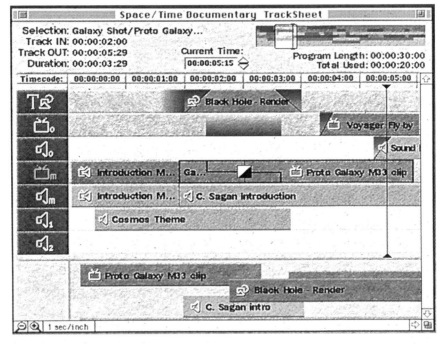

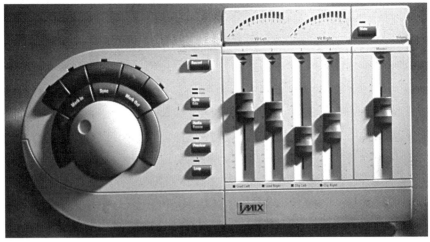

Manufacturer:	Integrated Research
Product:	Integrated Video, Harmony
Introduced:	IV: Siggraph 1992; Harmony: Siggraph 1995
Address:	2716 Erie Ave., Suite 2-W, Cincinnati, OH 45208
Phone:	(513) 321-8644 *Fax:* (513) 321-8722
Platform:	SGI

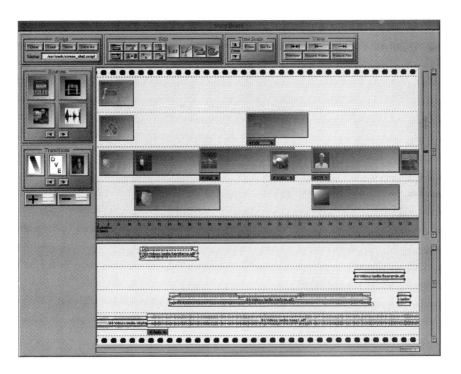

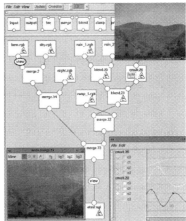

Harmony video production software (above) with a detail from the image compositing window (right).

Manufacturer:	in:sync
Product:	RAZOR
Introduced:	NAB 1993
Address:	6106 MacArthur Blvd., Bethesda, MD 20816
Phone:	(301) 320-0220 *Fax:* (301) 320-0335
Platform:	PC

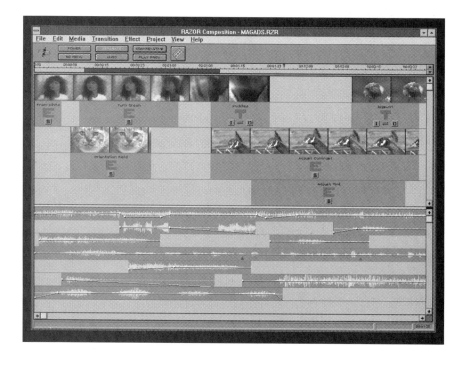

Manufacturer:	Lightworks UK, a division of Tektronix
Product:	Lightworks, Heavyworks, NewsWorks
Introduced:	NAB 1991
Address:	6762 Lexington Ave., Los Angeles, CA 90038
Phone:	(213) 465-2002 *Fax:* (213) 463-1209
Platform:	Proprietary PC

Photo courtesy of Lightworks

Manufacturer: Matrox Electronics Systems
Product: Matrox Studio
Introduced: 1991
Address: 1055 St. Regis Blvd. Dorval, Quebec, Canada H9P 2T4
Phone: (514) 685-2630   *Fax:* (514) 685-2853
Platform: PC

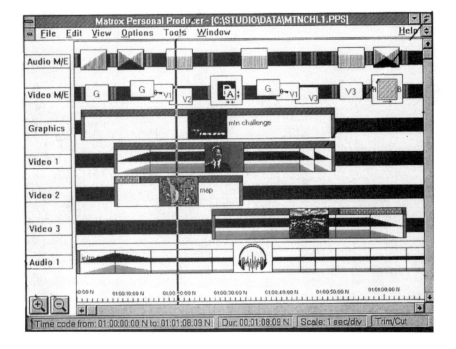

Manufacturer:	The Montage Group, Ltd.
Location:	Boston, Massachusetts
Product:	Montage III, Personal Personal Processor (MP3)
Introduced:	Montage I: NAB 1984
Address:	4116 West Magnolia Blvd., Burbank, CA 91505
Phone:	(818) 955-8801 *Fax:* (818) 955-8808
Platform:	PC

(special versions mfg'd for: BTS Bravo VE Virtual Control Editing System and MediaPoole and the DEC AlphaStudio)

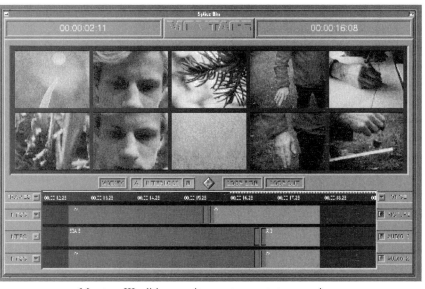

Montage III editing monitor, system prototype version

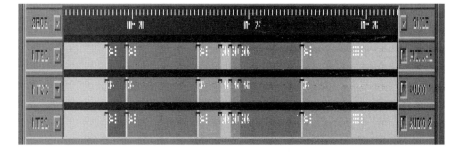

One of many views an editor can see of the bin — a kind of timeline for the cut sequence.

Manufacturer:	Panasonic
Product:	The Post Box
Introduced:	NAB 1995
Address:	One Panasoinc Way, 3F-5, Seacaucus, NJ 07094
Phone:	(contact regional Panasonic distributor)
Platform:	Proprietary PC

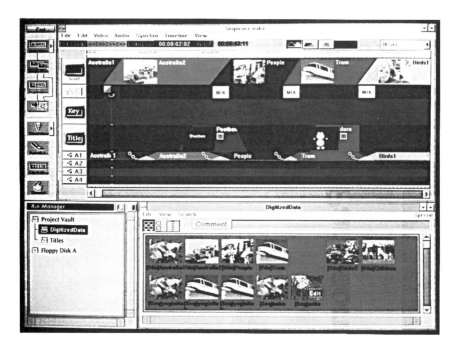

Manufacturer:	Play, Inc.
Product:	Trinity Preditor
Introduced:	prototype NAB 1995; release 1996
Address:	2890 Kilgore Road, Rancho Cordova, CA 95670-6133
Phone:	(800) 306-PLAY *Fax:* (916) 851-0801
Platform:	PC

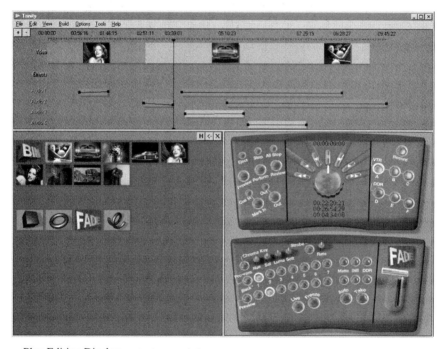

Play Editing Display — system prototype

Manufacturer:	Quantel
Product:	Edit Box
Introduced:	1995
Address:	(US) 85 Old Kings Hwy, North, Darien, CT 06820
Phone:	(203) 656-3100 *Fax:* (203) 656-3459
Platform:	Proprietary

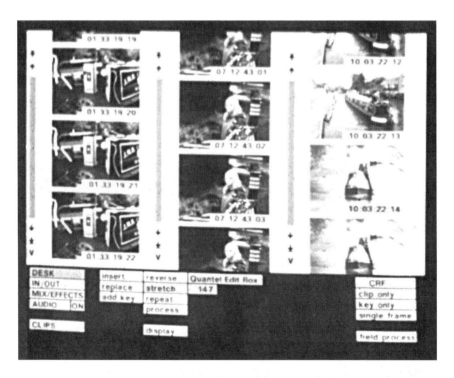

Manufacturer:	Ulead System
Product:	Media Studio
Introduced:	1994
Address:	970 W. 190th St., Torrance, CA 90502
Phone:	(310) 523-9393 *Fax:* (310) 523-9399
Platform:	PC

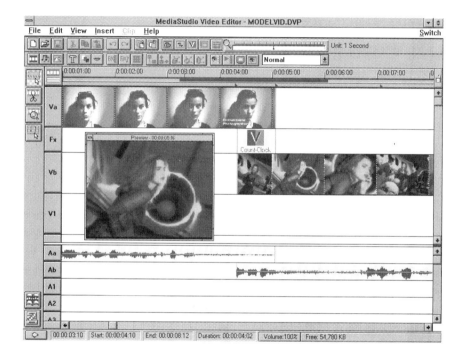

THE BEST SYSTEM

Bet you turned here first.

Okay. This is a trick, really. As you may have suspected, I'm not going to tell you what I think is the best system. Not because I think you might not agree, but because. . . there is no "best" nonlinear editing system. The question is actually kind of ridiculous. What do you mean by "the best"?

Least expensive? Best buy? Most versatile? Best for commercials? For drama? For theatrical films? Easiest to learn? Most popular? Most powerful? Most editors trained on it? Fastest cutter? Most reliable? Most likely to succeed? Best design? Most fun to use?

At least for these specific questions I have my own personal answers. But don't let anyone try to tell you which is best. Editors have many creative and technical preferences and vary based on their individual style of working and the types of projects they cut. Some editors swear by uprights; other editors prefer flatbeds. Every system — film, videotape, electronic nonlinear — every editing system has fans and proponents, has owners making money, has clients totally satisfied, has success stories, has lists of credits. And of course, every system has detractors and drawbacks.

A few years following the introduction of a new system, the big features are pretty much working well, and the little problems are mostly hammered out. The systems by then are usually very stable, running well-tested software, and have a pretty large trained base of editors. The manuals are usually done by then, and customers are pretty happy.

If an editing system works — no matter how ridiculous its interface, cumbersome its editing style, or incredible its competition — if it works, someone will use it, and someone will love it.

If you have looked around, tried a number of systems, talked with working editors and up-to-date facilities, and have located one system you like to use: you've done it. You've found the *best* system. Congratulations.

CHAPTER 6

THEORY

HORIZONTAL and VERTICAL NONLINEARITY

For this book, we define nonlinearity very specifically; in summary, it is a kind of perpetual preview. But let us delve further into the idea of nonlinearity, for it is central to understanding technological change.

An editing system is "nonlinear" as long as edits are not recorded to a master record videotape. Once shots are recorded from source to master, they lose their individuality, their distinctness. As long as shots remain distinct, it is quick and easy to manipulate them individually. Here are four shots, each from a different source reel:

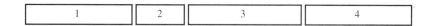

As long as these shots remain separate, and can be previewed, I would say they are "nonlinear." I would go farther as to define this separation as HORIZONTAL separation and the preview of these discrete elements is HORIZONTAL nonlinearity. They are horizontal in this graphic; horizontal because they occur consecutively in time. Horizontal nonlinearity is concerned with *temporal* separation.

If I were to record this nonlinear sequence to a master tape, it would look the same, but I would have lost the horizontal separation: they would now exist as a single entity. Traditionally in the video domain this action is called an "assembly." Let us define the loss of any separation as an *assembly*.

Manipulations and modifications in the pure horizontal dimension are very facile and require little mental or computational work. Like shuffling cards, or trimming shots, or re-arranging slides, I can choose when each shot begins and ends, and the order in which they fall. I need to put shots in, take shots out, and ripple or not. I would go so far as to say that pure horizontal manipulation is mechanically simple. Unfortunately, there is little "real world" editing that looks like this.

Now, if time is frozen, and we are dealing with a single frame, there are still manipulations that go on.

Think of the final product: at any given instant, there is more than the single frame of picture — there is sound, and usually many tracks of sound. Often there are graphics: titles, effects, and so on. When these simultaneous events are depicted graphically, the most common method is by showing layers of tracks:

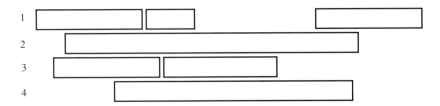

These layers show when events happen in relation to each other. These are vertical layers. As long as these vertical layers are maintained separately, let us define this as VERTICAL separation; if they can be previewed, let us define this as VERTICAL nonlinearity.

Vertical nonlinearity is not concerned with temporal issues, but with spacial issues. Like horizontal nonlinearity, vertical nonlinearity is not specifically for pictures, but for both visual and auditory information.

Vertical nonlinearity, like horizontal nonlinearity, exists for as long as the vertical elements are maintained as separate and discrete pieces. As long as vertical elements are maintained separately, they can be easily manipulated and modified. In the audio domain, a loss of vertical separation is referred to as a "mix-down"; in the video domain, the loss of vertical separation is sometimes called "rendering"; for our purposes here, let us continue to define a loss of any separation as an *assembly*.

It should be noted then that *assemblies, mix-downs,* and *rendering* are essentially the same kinds of processes — they combine separate elements into a less complex object — in doing so, separation is lost and nonlinearity is removed. And in all cases, this loss of separation is performed in order to view or preview a work.

It may be taken as fact that resources (energy, money, storage, *etc.*) must be utilized to maintain vertical or horizontal separation. Resources are saved as separation is lost. Ideally, elements would always exist with full separation, but reality dictates that at some point in the creative process, assembly is required. For the most part, even if separation can be economically maintained throughout the creative process, delivery of completed projects tends to follow a vertical or horizontal assembly.

The concept of synchronization, or "sync", is only fundamental to vertical processes. Clearly, editing as we know it today is a combination of vertical and horizontal manipulations. Viewing the interim stages or completed work is only possible when technology can simulate the assembly process. From a technology standpoint, it is easier to maintain separation of horizontal video elements and preview them (the common description of "nonlinear editing") than it is to maintain separation of vertical video elements and preview them (commonly called "real-time compositing"). But one can certainly assume that vertical nonlinearity will continue to evolve and proliferate as computational power increases.

Videotape editing, for years, was the editing of picture and two channels of sound, with three vertically separate tracks, but no horizontal separation except during the preview. Newer formats of tape made four channels of sound feasible. Additional or complex sound work is traditionally done on equipment designed to maintain vertical separation. A 24-track audio recorder does this. A tapeless digital audio workstation (DAW) uses hard disks to create a horizontal nonlinear system (remember that vertical separation existed prior to the "tapeless studio").

Effects and graphics have for years been synthesized and previewed in digital and analog domains, with some vertical and horizontal nonlinearity, but then mixed down and assembled to videotape.

As nonlinear editing systems evolved for picture, their evolution was primarily in the horizontal dimension. The unit of nonlinearity was the "load." A load consisted of a set of analog videotapes or videodiscs, and was limited. When the load became digital, on hard disks or MO disks, resource allocation could radically change.

Digital storage brings up the issue of *reciprocity:* As some factors go up, others go down. Most commonly, as image quality (color, resolution, size) goes up, hours of storage goes down. If the goal is to store many hours of source material on-line simultaneously (*i.e.* to build larger "loads") then clearly, you need to reduce image quality to meet this goal. This can be graphically shown; with image resolution (representing the quality of the image) on the vertical axis and quantity of source material on the horizontal axis, a "utility" line crossing both axis sets up the inverse relationship.

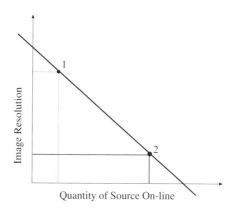

At point 1 the image resolution is high and thus the quantity of source on-line is low; conversely at point 2 the on-line quantity is large and thus the image resolution is low.

To make this graph more realistic, we can add some scales to the axis: for the image quality (y-axis) we can apply some generally understood degrees of quality; on the on-line source axis (x-axis) we can use a scale where each increment doubles the source of the prior value, beginning with 11 minutes of source and ending with 100 hours.

We will come back to this scaled graph in a moment.

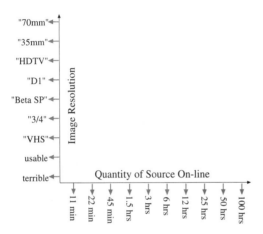

It should be noted that as time and technology march forward, the graphical result of the increases in computer functionality will shift the utility line outward. This includes better and faster CPUs, larger and faster hard disks, faster bus and communications hardware, improvements in video and audio compression and so forth. There are many elements that are limiting factors in the evolution of computers in relation to these graphs; for example: larger hard disks wouldn't alone allow for better image resolution while they may allow for more source on-line; they are "bottle-necked" by hard disk data transfer rates. Similarly, once data transfer rates improve, and compression makes better images fit in smaller files, other factors (*e.g.* the power of the CPU) may dictate the limit on how nonlinear (previewable) composited images may be.

This brings up two points. I would suggest that vertical and horizontal separation and nonlinearity are also inversely proportional. That is, with a fixed set of resources, to increase vertical separation reduces horizontal separation, and vica versa.

While from a technology standpoint this inverse relationship may not be appreciable due to the relative simplicity of horizontal separation when

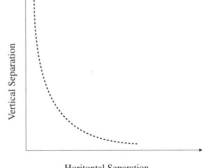

Horitontal Separation

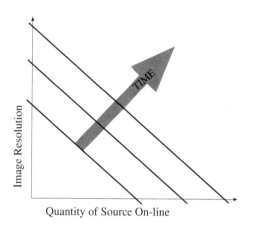

Quantity of Source On-line

compared to vertical separation, I might go on to suggest that while the toolsets required for horizontal and vertical nonlinearity might be able to be codified and integrated due to many of their inherent commonalities, the tasks related to horizontal and vertical manipulations are mutually exclusive: the more vertical layers maintained separately, the more difficult any single horizontal manipulation will be. This is why traditional work in the vertical domain is usually restricted until after (or later) in the horizontal process. To put it simply, sound is done after picture for a reason. If you work a dozen tracks of sound and then recut the picture, it is more work than waiting to do the sound after picture is locked. The reality of film making and post-production tends to push this as far as it can go, but ideally, vertical work — both for video effects and audio — is delayed until after picture is done.

Let us return to the evolution of computers, then, and the idea of computer complexity. In this case, complexity* is an abstract concept that ties together issues of computational complexity, required data bandwidth and transfer rate, and required storage. These features evolve at their own rate, but the net effect of this evolution is that different tasks become viable using computers of a given cost, with less complex ones first and more complex ones later. A simple diagram illustrates the point:

It is the complexity of computers that forms the bottleneck in the outward movement of the utility line on the earlier graphs.

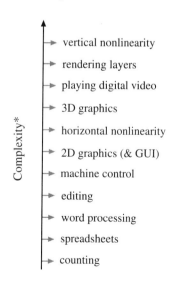

When we combine all of these observations, and look at the realities of post-production, the trends of the past decade are clear and a direction may become illuminated. First let's look at a set of typical post-production projects in terms of these values. Starting with quantity of source on-line, we can simply plot roughly how much source material is required for each type of production:

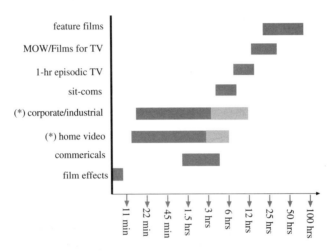

From the graph, we can see that to perform film effects we only need a few minutes of source material; to edit 4-camera TV sitcoms we need between 6 and 8 hours on-line; to cut a feature film we need between 24 and perhaps 100 hours on-line; and so on. So to utilize a nonlinear system for one of these types of productions, we now know the limiting value on the horizontal axis. For example, if the system can manage a maximum of 10 hours of source material, it will be difficult to edit a feature film, just about right (but on the low side) to cut an episodic TV show, and more than sufficient for most commercials and home videos.

We can do the same thing for the vertical axis, image resolution. The image resolution of broadcast TV is less than that for feature films but much higher than that required for home videos or many corporate projects. Does this mean that you cannot edit a feature film on a system that only has the image quality suffi-

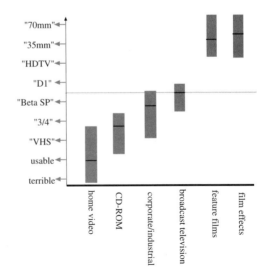

*Note that these source requirements are general figures, and for many projects, especially consumer or corporate, the values have enormous variations.

cient for home videos? No, but you cannot deliver your product directly from the system. When a system's output image quality is less than that which is required for delivery, you have the natural division between online systems and offline systems.

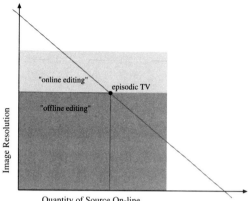

Clearly, online systems are desirable in that they will require no other steps to deliver product; however, as long as economics prohibit the proliferation of online systems, offline may have to be sufficient. Affordability determines system complexity, and affordability is itself largely a function of prior expecta-

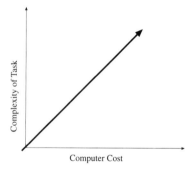

tions. TV post-production *expects* to pay thousands of dollars per week for equipment to edit; as time goes on, those productions have a choice — get more functionality for their money, or get the same functionality for less money.

What is happening in our post-production market is that as the line on the first graph shifts outward due to technological improvement, it will continue to move first in the domain that is most required until the maximum degree of value is reached, and then all future shifts in the line are immediately translated into increases in the less-required domain.

For example: feature film editing systems. They require a large quantity of source material (most-required). They will take whatever image resolution they can get. At first, image resolution is sacrificed to make sure all source can be loaded. But once that maximum threshold is reached (let's say 50 hours of source), additional improvements in compression and hard disk storage will be shifted towards improving the image quality. Since the *status quo* of film editing is offline, there is not a great deal of discussion about the reality of online film editing. However, you can be assured that at some point in the future, when film-quality digital video can be stored on a system costing a few thousand dollars per week, and the quantity of source hits the

18-25 hour range of our graph, people will want to print their digital video directly to film.

We can surmise that there will be a point in time prior to the above scenario where "broadcast-quality" video (lower quality than feature film) can be managed on a several thousand dollar a week system and the source on-line begins to rise. At first, short-form projects will utilize these systems for delivery: commericals, then longer projects like episodic TV. This point has already come, as would have been predicted, for news, home video and corporate video.

This discussion is oversimplified, but it should illustrate the kinds of issues in the evolution of nonlinear systems and nonlinear online.

FACETS OF POST-PRODUCTION TASKS

Horizontal tasks and vertical tasks are each unique, and perhaps ideally performed with specialized tools. It is certainly clear that issues of efficiency are often best handled by the separation of labors. To put this is today's terminology, I would say that *traditional* offline editing is mostly a horizontal task, best performed with horizontal tools — it is the building of the story, of the framework of a production, it is the process of culling down large volumes of source material into a small volume of relevant material; whereas *traditional* online editing is mostly a vertical task, performed best with vertical tools — it is the integration stage, it begins with a locked picture and then augments the look of the elements and integrates the cut with other elements, like titles and special effects.

Vertical systems, tailored for vertical separation and manipulation, are generally not well designed for "editing." Nor should they be. Horizontal systems, on the other hand, are only slowed down and complexified by the addition of many vertical elements (either for picture or sound); here it is perhaps most efficient when the verticality is kept to a minimum.

Equipment tends to have a strong bias towards either vertical nonlinearity or horizontal nonlinearity. Note that horizontal and vertical distinctions have no direct relationship to online and offline systems. Recall that online and offline are relative terms related to output **product**: *online quality* can be defined as the quality required for actual product delivery; *offline quality* is any image or sound quality less than that. There is a tendency to relate vertically nonlinear systems with online quality and horizontal nonlinear systems with offline quality. Although this is a fair observation, it is important to understand that it is coincidental; because many complex effects are difficult to "describe" for re-creation, they tend to reside solely in the online domain, and thus verticality coincides with online quality.

Vertical and horizontal functionalities are not the only manipulations in today's systems. Another functionality is what I call *SYNTHESIS*. Graphically, it occupies no space, but produces the elements that are to be manipulated in the two dimensions. Shooting on location is a synthesis function (and is not readily managed on a computer). Creating 3D graphics is synthesis, as are 2D painting and text

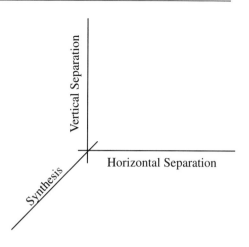

elements. Every system produced balances the functions of horizontal, vertical, and synthesis as required by the market for that product.

Not only can object-frames be created and maintained as separate frames vertically and/or horizontally (inter-frame), but synthesis functionality can also involve objects *within* a frame (intra-frame). The intra-frame objects — like object-oriented graphics (PostScript) and bit-mapped objects — also can be manipulated with horizontal and vertical nonlinearity, *i.e.* they can be stacked and blended (vertical), and moved around the screen (horizontal). A film or videotape camera creates frame-objects of a single vertical layer, and at the intra-frame level are like bit-mapped graphics.

The new trend in nonlinear online systems is products that have vertical nonlinearity as well as horizontal nonlinearity, limited source capacity, and some synthesis functionality. More traditional nonlinear systems like the *Lightworks* are extensively horizontal, with limited verticality, generally offline quality, and negligible synthesis functionality. This makes sense for the narrative (horizontal) markets for which the product was originally designed. Discreet Logic's *Flame* is a very vertically nonlinear system with limited horizontal functionality (roughly corresponding to its limited source capacity), but with considerable synthesis capabilities. Consumer-oriented products have a healthy mix of vertical separation (although complexity limits vertical nonlinearity), horizontal nonlinearity (the functionality is easy to deliver, especially at lesser image resolutions), and synthesis. More often than not, however, consumers and some professionals utilize the comprehensive synthesis functionality of products like Adobe's *Photoshop* and *Illustrator*, Macromedia *Freehand*, and Fractal Design's *Painter* — and simply import objects/frames into horizontal or vertical systems.

These theoretical constructs are not exclusive to video. The much-discussed desktop publishing realm is analogous as well. Writing and word processing are horizontal tasks, and we see proliferation of dedicated tools for the process. Publishing, on the other hand, is managed by specialized publishing tools. It is the integration and synthesis stage that begins with the locked manuscript, and then continues from there. The online *assembly* in the publishing industry is called printing. The real-time *preview* that characterizes nonlinear systems is generally referred to as WYSIWYG ("what you see is what you get").

In fact, with very little difficulty the tasks of publishing and video can be integrated where frames equal pages, TV safe lines equal margins, objects are imported, and output options are varied. While this integration may not yet have been perfected, this is the essence of multimedia authoring software.

There is no question that word processing software has many publishing functionalities; and publishing software has ample word processing functionalities; and yet the two tasks remain distinct, even by those who perform both. The phenomenal advent of desktop publishing, with its associated software products, has not eroded the word processing software market — in fact, reports seem to indicate that it has increased it. Professional writers and authors still tend to use the least complicated, most streamlined software for word processing.

This is not meant to indicate that traditional "horizontal" editors do not desire the additional power associated with verticality or with synthesis, only that they will desire it to the extent that it does not interfere with the efficient completion of their task.

I would suggest that the ideal human interface for horizontal manipulations is one that embodies rhythm and motion and is very physical, like a dedicated console. Vertical manipulations are perhaps best managed via graphical representations on computer displays and a corresponding mouse (or similar device). Can a graphical display manage horizontal elements? Absolutely. Can a physical console manage vertical elements? Somewhat, but perhaps to a lesser degree.

The bottom line is that even as traditional separations of labor — like online or offline — begin to erode, there may still be other rationale for a division of labor; the tasks associated with each arm (horizontal, vertical, synthesis) are specific and tools dedicated to those tasks will continue to thrive in the professional arena over the next many years. For consumers, generalization will be the trend, as systems go for high-profile functionality and diverse needs of a huge market, with less sensitivity to the subtle functions and styles of dedicated products. As will all products in all industries, there will be a continuum of product functionalities from the general (consumer) to the specific (professional).

THE FUTURE OF OFFLINE
An Interview with the Author

Is offline dead?

Good question. Personal computers can presently manipulate some pretty good video. And many projects — beginning with home videos, then in-house corporate and training videos, and moving slowly up the dollar ladder towards professional broadcasting — will be able to utilize this output. **Thus, you are online!** Already high-end nonlinear editing systems have delivered output *directly to broadcast air.* It doesn't get more online than that. And as the delivery quality improves at affordable economics, more and more applications that run on personal computers will control some video decks, perform fancy titling and graphics and manipulations, do multi-track audio work. . .

Providing you know what you are doing.

However, with that having been said, I'd still reply "Offline is doomed, but not quite dead. Not for a while." Because after all is said and done, the defining characteristic of offline is that it is cheap — by definition, **cheaper than online**. So no matter how cheap online gets (and it WILL get cheaper), if I can do *almost* the same thing for LESS money, there will be some kind of offline. Even if your fancy fully featured online bay of tomorrow only rents out for $50/hour (today they might be $500/hour) then I can probably rent a cheaper computer to do non-online quality stuff for $5/hour. I can spend 10 times as long playing around with my material for the same cost, and only go to "online" to assemble the final elements. And I will have motivation to work this way as long as the transfer of edit and graphic decision information is not too much trouble. Standards like Avid's OMF Interchange may in some ways extend the life of offline.

The death of offline will be marked both by the major drop in the costs of online quality equipment *and* when the types of effects you can do on a computer defy easy translation; it will simply be easier to do it all at one system. Depending on the type of product you want to produce, offline may be around for quite some time. Feature films, on one hand, will remain a divided task for many years; on the other hand, corporate video is in some ways beginning to end its relationship with offline.

Wow. $5/hour? You think offline will be that cheap?

No. I was just making a point. Look at desktop publishing. It's cheap and available to lots of people today. I myself am fanatical about it. But to do it right, it's still marginally expensive — what with the laserprinters and scanners and big monitors and so on. Right now, I'm holding in my hand a price sheet from a local self-service Mac rental place. Something we might see more of in the future — you walk in, rent their hardware and software for a few minutes or hours, use their high quality printers — their price for a Quadra 700 with a 20" monitor, 24-bit color, Syquest Drive, and color scanner is $22/hour. This is still not a sufficiently configured system to do any video — online or offline. So the fact is, I don't know. But however it is priced, offline will ultimately be defined by its significant price savings over online.

It is also very important to remember that the old video equipment lasted pretty long. Even if better equipment was created, your old stuff still generated a fair amount of money. In business terms, you could amortize your purchases over many years and thus rent it economically. Computer equipment, even if it cost exactly the same as the traditional video equipment, is rendered obsolete very quickly — perhaps quicker than any industry has ever had to deal with — and so the amortization of that cost is only allowed over a year or two. This keeps costs higher, and makes investments a little riskier.

I would also wager a guess that if my first task is to cull down 10 hours of video into a 15 minute sequence, I really only need to pay for something to do just that, and do it pretty quick and easy. To pay even a few cents more for a huge array of effects and audio and such just may not be worth it — not until later.

You don't think people want all that additional functionality for a small additional cost?

Many will. But if offline becomes the pure domain of the "edit" where online is everything else (which is what I think will happen), doing too much work in offline is somewhat counter-productive. Today I hear reports from online houses getting the output (lists and tapes) from the high-powered nonlinear offline systems. The cuts are great, the offline editors have done lots of audio work and many effects, but when the actual elements are going together, the client wants a different effect (from a fancier box) or wants something pulled up a few frames at the last minute, and the digital audio output from the offline system is no longer right, or the effects need to be re-worked. . . .

I'm only saying that offline will also be a place to sell the cut, with perhaps as little work as possible. Sometimes it doesn't pay to get everything just perfect before you're working with the final elements. Things change.

You sound defensive. Why?

I sound defensive? Hmm. . . I think that many people are being shown a rosy picture of desktop video and given a lot of promises. I love it myself, but the political and economic ramifications are beyond anyone's current estimations, and I think computer companies would like us all to believe it's a sure thing and present tense (as opposed to a likely future). The fact is, a lot of work needs to be done today, and much of the new promised technology doesn't really solve today's problems.

So the question is, when does the future start?

Nice question. I suppose someone with vision has to buy new equipment and experiment and wait it out and drive the development even if it doesn't really makes sense to the mainstream. That's what is going on today. It will be a few years before desktop video solves more problems than it creates. Again, look at desktop publishing. It is about 10 years ahead of desktop video but still is not even close to a "no-brainer." This is partly due to the burden of using high technology to produce a low-technology product (there is much confusion created in the interface between the electronic and print world); however those issues aside, desktop publishing still requires some phenomenally expensive equipment at certain parts of the process and a vast array of skills beyond the obvious electronic pasteboards for layout, graphics and writing. Desktop video, I presume, will allow more people to enter the video realm, but will still take some significant equipment and know-how to deliver high-quality product. But the pioneers are out there now, figuring it all out, and defining the stuff in a way that will make sense to the rest of the world.

Is tape dead?

No. A simple physical fact of life is that no matter how much data you can store on a disk, I can store more, for a lot less, on a linear tape. Tape is an excellent archival medium. Its biggest drawback, as we all know, is that it is linear, and thus not particularly random access. We need to ask ourselves how much people are willing to pay for random access. (It's also worth noting that there can be semi-nonlinear systems that are not random access.)

If you can buy a 1GB drive for $400 and I can buy a 2GB digital tape for $10 dollars, well, heck, if I am willing to wait just a bit for some cueing, I save a bundle. Is there a place for that in the world? I would guess yes. At least until some radical new memory device rears its head and changes everything — and a radical new fast giant cheap memory device would CHANGE EVERY-THING. Don't think someone somewhere isn't working on it.

Another point is that there are tons and tons of tapes and videotape decks out there working today. Even if someone invented something inescapably perfect today, everyone is not about to throw out all the working stuff they already own. At least not yet. The real change occurs when people STOP purchasing the old stuff. When old stuff breaks and they replace it with the perfect new stuff: that's when the revolution begins. Today, the stirrings are there, but at present, most people who purchase nonlinear equipment are augmenting their facilities, not removing existing bays (and professionals are not only arguing about nonlinear online, but also "should I get D5 or

DCT?"); there is still a great deal of linear going on, being purchased, lots of tape decks moving, lots of effects boxes being sold. And with HDTV around some corner, perhaps tape controllers aren't such a bad idea. I would have to say that the revolution, which is bound to happen, is beginning the way many revolutions begin: with lots of confusion and instability.

But we can see it from here.

Last question: What is going to happen to all the digital nonlinear systems out there?

You mean, the proliferation of products? What's going to happen is perhaps what is currently happening: more and more products besiege the market... computer companies trying to cash in on the rich video business, video equipment companies upgrading their product lines and client base with better technology... a new generation of lower-end video people doing whatever they can do on their PCs with their camcorders. . . mostly just for fun. . . .

I think there will be some serious market niche stuff going on. Certain products just happen to be better for certain kinds of projects — "better" meaning faster, easier, more appropriate, more economical. I think that systems will cease to be plain old editing systems, and start to be more "commercial editing systems" and "home computer video workstations" and so on. In fact we've already seen dedicated "news cutters" and "feature film offline" systems. This will continue.

Certainly, a motivated user can make any system do exactly what he or she wants — you can cut a movie today on a home VCR, if you were so inclined. It might be a tedious experience, but if you were up for it, and had the equipment, who could beat it?

If current indications tell you anything, networking will be the future of offline: offline systems will simply be connected to the online systems and the servers with high-quality images. Magazine publishing has been doing this for years: everyone uses workstations doing layout and writing; the images are "pasted up" and such at great resolutions, but not the resolution required for the printing device. *Those* online images are only downloaded and stripped in at the end. This seems to be where high-end "desktop publishing" has gone, and it does look like a plausible methodology for digital video editing in the near future.

I think professionals will always demand high-quality, finely tuned, robust products. Period. I think everyone else may just get caught up in the craze and buy whatever the market deems popular. Many thousands of systems will be sold to consumers and prosumers. A few thousand systems will satisfy the professionals. The professional market may be too small to attract the bulk of the computer industry's attention.

Anyway, that's my guess.

Appendix A
METRIC NUMBERS

If you didn't get swept up in the metric reforms of the 70s, you may be lacking an essential knowledge of metric number prefixes. Metric numbers are based on the number ten (10). They were devised to make doing math and science particularly easy. For the uninitiated, here is the key:

1,000 (thousand)	=	kilo-	(K)
1,000,000 (million)	=	mega-	(M)
1,000,000,000 (billion)	=	giga-	(G)
1,000,000,000,000 (trillion)	=	tera-	(T)

By using these prefixes, you can save yourself the writing out of a lot of long numbers. For example, rather than saying:

1,000,000,000 bytes of computer data*, you say 1GB (one gigabyte).

1,000,000 cycles per second (Hertz), you say 1MHz (one megahertz).

1,000 grams, you say 1Kg (one kilogram).

Just as people often abbreviate one kilogram as "one kilo," so do engineers and computer users shorten megabytes and gigabytes to "meg" and "gig." It's a little strange if you're not used to it, but it's actually quite simple.

* IMPORTANT NOTE ABOUT METRIC NUMBERS *
AND COMPUTERS

The above table accurately reflects the nature of metric numbers; unfortunately, the computer business has quietly modified one of these rules to meld with its own reality.

A thousand *somethings* is supposed to be a kilo-*something*, except when it comes to **computer bytes**. Because computers count with binary numbers, a computer that counts to 1000 can also count to 2^{10}, or 1024, and therefore, a kilobyte is *not* 1000 bytes but actually **1024** bytes.

Thus: a kilobyte (KB) is 1024 bytes, a megabyte (MB) is 1024 kilobytes, and a gigabyte (GB) is 1024 megabytes.

FILM FOOTAGES

Just as metric numbers are "base 10" and binary numbers are "base 2," in a way, film numbers are "base 16." As you count, when you get to 15 you roll over another foot. Here is a small table that may help videotape editors and others unfamiliar with film footage counts:

For 35mm film, 4 perfs per frame...

1 foot	=	16 frames	=	2/3	second
1' + 8	=	24 fs	=	1	second
6'	=			4	seconds
15'	=	240 fs	=	10	seconds
30'	=			20	seconds
45'	=			30	seconds
90'	=	1440 fs	=	1	minute
180'	=			2	minutes
450'	=			5	minutes
900'	=			10	minutes

When counting film feet and frames, you must predetermine whether the first frame you start on is number "0" counting up to "15" or if the first frame is number "1" counting to "16." Zero counting (0-15) is the most common numbering scheme. You also must decide where you define your "0" frame in relation to each key/code number.

Kodak's introduction of KeyKode®, their machine-readable edge numbers, solves these issues by standardizing and marking each 'zero frame'. Adoption of this format will make the reading of edge numbers considerably more simple for all film handlers.

BINARY COUNTING

Binary mathematics is a kind of counting in base 2 (it's no different from base 10, if you're missing eight fingers):

Decimal		Binary			
Place 10	Place 1	Place 8	Place 4	Place 2	Place 1
	0				0
	1				1
	2			1	0
	3			1	1
	4		1	0	0
	5		1	0	1
	6		1	1	0
	7		1	1	1
	8	1	0	0	0
	9	1	0	0	1
1	0	1	0	1	0

Just as counting in base 10 is based on powers of ten, base 2 is based on powers of two. The powers of two, listed here, are familiar multiples to computers users:

$$2^0 = 1 = 0 \text{ bits}$$
$$2^1 = 2 = 1 \text{ bit}$$
$$2^2 = 4 = 2 \text{ bits}$$
$$2^3 = 8 = 3 \text{ bits}$$
$$2^4 = 16 = 4 \text{ bits}$$
$$2^5 = 32 = 5 \text{ bits}$$
$$2^6 = 64 = 6 \text{ bits}$$
$$2^7 = 128 = 7 \text{ bits}$$
$$2^8 = 256 = 8 \text{ bits}$$
$$2^9 = 512 = 9 \text{ bits}$$
$$2^{10} = 1024 = 10 \text{ bits}$$

2^{11}	=	2048	=	11 bits
2^{12}	=	4096	=	12 bits
2^{13}	=	8192	=	13 bits
2^{14}	=	16384	=	14 bits
2^{15}	=	32768	=	15 bits
2^{16}	=	65536	=	16 bits
2^{17}	=	131072	=	17 bits
2^{18}	=	262144	=	18 bits
2^{19}	=	524288	=	19 bits
2^{20}	=	1048576	=	20 bits

The number of bits is the number of places a digit can have:

In 8-bits, the smallest number is 00000000 = 0
and the largest number is 11111111 = 255

The multiplication of two binary numbers always produces a result that has a total number of digits equal to the sum of the digits of the original numbers; for example:

10010 x 011110 = 1001010100
(a 5 bit number times a 6 bit number equals an 11 bit number)

This might help explain why computers need many bits to do binary math — for digital audio and video manipulation (like 3-D graphics or music synthesis, *etc.*) Doing these computations in "real time" means a computer is doing millions of computations a second and moving large numbers (many bits) around the computer quickly. This takes more expensive computers with larger chips and faster processors.

For the digitizing of video, every pixel on the screen can have any one of a number of colors associated with it — the more choices of colors you have, the better the approximation of "reality." The number of choices for color are usually represented by the number of bits each pixel will be stored in. For example, a 16-bit system will allow each pixel to be one of 65,536 colors.

The greater the color choices, the better the images, but also the more memory needed to store an image. Many people consider "film-quality" color to be 8 bits for each of red, green and blue — or 24 bits total, or 16.78 *million* colors. The same image stored with 24 bits rather than 8 bits takes up more than 65,000 times as much memory.

THE ACADEMY AWARDS

Awards for Best Editor have been given out by the Academy of Motion Pictures Arts and Sciences since 1934. The winner is not always from the same film that wins for best picture, although it is somewhat arguable that a "best picture" owes a great deal to a good editor. It is also difficult to imagine how a person wins an award for "best" editing while the films are only being judged by the final product.

Without knowledge of the conditions from which the film was derived —quality and quantity of dailies, time constraints, personalities involved— it is hard to compare two films' editing. One editor could have taken a ridiculous story with bad footage and turned it into something wonderful; another editor could have had a clear story from a director with a good vision, exceptional footage, and turn out an equally wonderful film. Who's the "best" editor? It is pretty well understood that this is a difficult judgment.

The list below began simply as suggestions for films to watch for truly great editing technique. Editors (video or film) who work in narrative material should have some heroes. The history of *film* editing should be (if it isn't already) understood and appreciated by those who have a purely video background. But aside from a few obvious gems, it is difficult to choose what this list should include.

Consequently, I gave up, and changed the "suggested viewing list" to the Best Editor winners of the Academy Award. Also included is the film winning Best Picture, as it is not always the same.

ACADEMY AWARD WINNERS

For each year, the first line is the winner for Best Editor, and the second line *(in italics)* is the winner for Best Picture. One entry indicates a double winner. Most Best Picture films were also nominated for Best Editor; however, those that were not are marked with an asterisk (*).

YEAR	FILM	EDITOR(S
1934	Eskimo	Conrad Nervig
1935	A Midsummer Night's Dream *Mutiny on the Bounty*	Ralph Dawson *Margaret Booth*

YEAR	FILM	EDITOR(S
1936	Anthony Adverse *The Great Zeigfeld*	Ralph Dawson *William Gray*
1937	Lost Horizon *The Life of Emile Zola**	Gene Havlick, Gene Milford *Warren Low*
1938	Adventures of Robin Hood *You Can't Take it With You*	Ralph Dawson *Gene Havlick*
1939	Gone With The Wind	Hal Kern, James Newcom
1940	North West Mounted Police *Rebecca*	Anne Bauchens *Hal Kern*
1941	Sergeant York *How Green Was My Valley*	William Holmes *James Clarke*
1942	The Pride of the Yankees *Mrs. Miniver*	Daniel Mandell *Harold Kress*
1943	Air Force *Casablanca*	George Amy *Owen Marks*
1944	Wilson *Going My Way*	Barbara McLean *Leroy Stone*
1945	National Velvet *The Lost Weekend*	Robert Kern *Doane Harrison*
1946	The Best Years of our Lives	Daniel Mandell
1947	Body and Soul *Gentleman's Agreement*	Francis Lyon, Robert Parrish *Harmon Jones*
1948	The Naked City *Hamlet**	Paul Weatherwax *Helga Cranston*
1949	Champion *All The King's Men*	Harry Gerstad *Robert Parrish, Al Clark*
1950	King Solomon's Mines *All About Eve*	Ralph Winters, Conrad Nervig *Barbara McLean*
1951	A Place in the Sun *An American in Paris*	William Hornbeck *Adrienne Fazan*
1952	High Noon *The Greatest Show on Earth*	Elmo Williams, Harry Gerstad *Anne Bauchens*
1953	From Here to Eternity	William Lyon
1954	On the Waterfront	Gene Milford

* Indicates a film that won for Best Picture but was not nominated for Best Editor.

YEAR	FILM	EDITOR(S
1955	Picnic *Marty**	Charles Nelson, William Lyon *(spvr) Allan Croslund Jr.*
1956	Around the World in 80 Days	Gene Ruggiero, Paul Weatherwax
1957	Bridge on the River Kwai	Peter Taylor
1958	Gigi	Adrienne Fazan
1959	Ben-Hur	Ralph Winters, John Dunning
1960	The Apartment	Daniel Mandell
1961	West Side Story	Thomas Stanford
1962	Lawrence of Arabia	Anne Coates
1963	How the West Was Won *Tom Jones**	Harold Kress *Tony Gibbs*
1964	Mary Poppins *My Fair Lady*	Cotton Warburton *William Zeigler*
1965	The Sound of Music	William Reynolds
1966	Grand Prix *A Man for All Seasons**	Fredric Steinkamp, Henry Berman, Steward Linder, Frank Santillo *Ralph Kemplen*
1967	In the Heat of the Night	Hal Ashby
1968	Bullitt *Oliver!*	Frank Keller *Ralph Kemplen*
1969	Z *Midnight Cowboy*	Francoise Bonnot *Hugh Robertson*
1970	Patton	Hugh Fowler
1971	The French Connection	Jerry Greenberg
1972	Cabaret *The Godfather*	David Bretherton *William Reynolds, Peter Zinner*
1973	The Sting	William Reynolds
1974	The Towering Inferno *The Godfather II**	Harold Kress, Carl Kress *Peter Zinner, Barry Maulkin,* *Richard Marks*
1975	Jaws *One Flew Over the Cuckoo's Nest*	Verna Fields *Richard Chew, Lynzee Klingman,* *Sheldon Kahn*

* Indicates a film that won for Best Picture but was not nominated for Best Editor.

YEAR	FILM	EDITOR(S
1976	Rocky	Richard Halsey, Scott Conrad
1977	Star Wars	Paul Hirsch , Marcia Lucas, Richard Chew
	*Annie Hall**	*Ralph Rosenblum*
1978	The Deer Hunter	Peter Zinner
1979	All That Jazz	Alan Heim
	Kramer vs Kramer	*Jerry Greenberg*
1980	Raging Bull	Thelma Schoonmaker
	*Ordinary People**	*Jeff Kanew*
1981	Raiders of the Lost Ark	Michael Kahn
	Chariots of Fire	*Terry Rawlings*
1982	Gandhi	John Bloom
1983	The Right Stuff	Glenn Farr, Lisa Fruchtman, Steve Rotter, Douglas Stewart, Tom Rolf
	Terms of Endearment	*Richard Marks*
1984	The Killing Fields	Jim Clark
	Amadeus	*Nena Danevic, Michael Chandler*
1985	Witness	Thom Noble
	*Out of Africa**	*Fredric Steinkamp, William Steinkamp Pembroke Herring, Sheldon Kahn*
1986	Platoon	Claire Simpson
1987	The Last Emperor	Gabriella Cristiani
1988	Who Framed Roger Rabbit	Arthur Schmidt
	*Rain Man**	*Stu Linder*
1989	Born on the Fourth of July	David Brenner, Joe Hutshing
	*Driving Miss Daisy**	*Mark Warner*
1990	Dances With Wolves	Neil Travis
1991	JFK	Joe Hutshing, Pietro Scalia
	Silence of the Lambs	*Craig McKay*
1992	Unforgiven	Joel Cox
1993	Schindler's List	Michael Kahn
1994	Forrest Gump	Arthur Schmidt

* Indicates a film that won for Best Picture but was not nominated for Best Editor.

M I N I - L E X I C O N

ACE: American Cinema Editors. The film editors honor society.

A-D: analog-to-digital conversion. Opposite of D-A.

ADR: Automatic Dialog Replacement — replacing the original dialog recorded on location with dialog that has perhaps been recorded under better conditions, like in a studio. Because of the way ADR was once recorded, it is also called "looping."

alphanumeric: refers to the 26 letters of the alphabet and the numbers 0 to 9 on the computer keyboard. A typical alphanumeric is "scene A45C", where the scene number is actually a combination of numbers and letters. Many computers have problems in the entering or correct sorting of typical production alphanumerics.

ASCII: American Standard Code for Information Interchange. A data coding standard usually for computers. There are ASCII (pronounced "ask-ey") codes for all keys of a computerized keyboard — which allows typed letters to translate into a universal symbolic language.

AVR: Avid Video Resolution. A commonly used method of describing video resolutions used in digitizing.

batch digitizing: to input a list of shots to a nonlinear system whereupon the system automatically cues a source tape and records the list of selected shots into the attached digital storage devices. It is a form of auto-assembly used to get pre-logged material digitized; as opposed to marking a shot and digitizing it then, or digitizing on-the-fly.

Beta SP: a 1/2" tape format with high quality component video and audio; short for "Betacam" SP. There is also Digital Betacam.

beta test: the initial field testing of new computer hardware and software, prior to an official public release. An *alpha* test or alpha version of something is the original in-house release of a new product.

bit map: a method of storing and displaying images where each pixel that composes the image is recorded, like a little picture; as opposed to *vector* (or *PostScript*).

BNC: a common type of terminal used at the ends of video (and digital audio) cables characterized by a *twist-release* connect/disconnect function.

buffer: a site for temporary storage — usually of data in computer memory. Used typically to store edit list, or digital video or audio information before being transferred to a more permanent disk-based storage.

bug: a malfunction in computer hardware or software.

CGI: Computer Generated Image. CGI is a rapidly growing domain of image synthesis, in particular, where photorealistic images can be incorporated into (or replace) live action images. CGI is now used with video compositing and digital retouching, matte painting, and other special effects. The fundamental tool of modern CGI is the 3D software package.

chrominance: the color portion of the broadcast video signal, relating to the hue and saturation of the image. Luminance draws the picture that chrominance paints.

CMYK: an abbreviation for cyan, magenta and yellow — the three subtractive primary colors (and complements to RGB); K is black. CMYK color is used primarily in color printing, where RGB is used in color light and video.

color correction: part of the post-production process that deals with the scene-by-scene fine-tuning of the color and density timing of the product of the on-line edit.

component video: a video signal path that maintains separation of the R, G and B colors for very high-quality recording. Separate cables carry the colors.

composite video: a more common video signal that encodes and decodes the R, G and B colors. A single cable can conduct composite video.

compression: the technology of taking large amounts of computer data (sometimes a digital video image) and carefully squeezing it down to a much smaller size for easier storage and manipulation.

crash: a failure of a computer system, usually due to an error on a hard disk or a software programming error.

cue: to look at a scene take or to set up for an edit.

cuts-only: an editing system or style with no effects such as dissolves, wipes, fades.

DAT: digital audio tape. A technology of audio recording; the small, high-quality (digital) tapes are an alternative to 1/4" tape, and are being used for location recording and other applications.

DAW: digital audio workstation.

default: When a computer chooses the most common (likely) option in a menu by placing a cursor on that option or a numeric value in a data field. Computers often have default settings for when the user opts not to select something specifically.

DF: drop-frame timecode

disc: a "laserdisc" or "videodisc"

disk: a computer-kind of disk — a floppy disk, hard disk or magneto-optical disk.

DVI: a series of proprietary chips from the Intel Corporation. DVI, for Digital Video Interactive, has its own type of compression for digital video.

EDL: edit decision list—a sequential list of edits stored either in computer memory, or as a printout, or on a disk.

fps: frames per second. Sometimes spoken as "fips."

generations: refers to the number of times a segment of video image or signal is copied. Analog video and audio quality deteriorates with each generation. Digital information does not suffer any generation loss.

hardware: computer components with a physical form; mechanical or electronic equipment, as opposed to software.

HDTV: High Definition Television

IC: short for Integrated Circuit, a computer chip.

I/O: computer slang for "input/output."

ISDN: Integrated Services Digital Network. A method for transmitting all types of digital data (video, audio, phone, *etc.*) in a single digital network. Bandwidth is 64Kbits/second, compared to T1 (1.5Mbits/second) and T3 (45Mbits/sec) lines.

jog: to view video by moving it 1 frame at a time either forward or backward.

JPEG: Joint Photographic Experts Group. An organization that set out to standardize computer methods for the digitizing of images. You'd need to be a computer video engineer to really follow this stuff.

kinescope: a film of videotape, the opposite of a telecine. Kinescopes were used in the 50s to save film versions of live television broadcasts, before the introduction of videotape.

LED: Light Emitting Diode. A computer part that usually makes up red displays.

LTC: longitudinal timecode — audio-encoded timecode. As opposed to VITC.

luminance: the black and white portion of the broadcast video signal. Technically, it is referred to as "Y"; it is considered the most important part of the signal for conveying picture information.

MO Disk: Magneto-Optical. A specific type of optical memory disk used to store computer data.

MPEG: Moving Picture Experts Group. An *inter*-frame image compression method tailored for moving images and sync sound. MPEG-1 and MPEG-2 are implementations of this scheme.

NAB: National Association of Broadcasters — the television/radio association working for the interests of the broadcast industry. The NAB exhibition, a showcase for new equipment, is usually held each April in Las Vegas.

Nagra: Brand of 1/4" audio tape recorder, used commonly on film productions. Some Nagras have timecode generators recording timecode on the center track of the tape which allow 1/4" tapes to be more easily synchronized with video or film.

NLES: **n**onlinear **e**diting **s**ystems; sometimes quaintly spoken as "nellies".

NTSC: National Television Standards Committee. The American organization that standardizes video signal characteristics. NTSC video is 29.97Hz (almost 30fps) at 525 horizontal lines per frame.

NuBus: the pathway of information around traditional Macintoshes; replaced by PCI bus (Peripheral Component Interface) which is more in the PC mainstream and has significantly higher bus width (bus or bandwidth is measured in MHz).

offline: (*traditional*) the creative editing of video using copies of master videotapes to create an edit decision list.

on-line: to be connected to a computer or device or system . . . as opposed to . . .

online: (*traditional*) the part of post-production where delivery-quality elements are created, integrated, and assembled. Online utilizes the offline EDL to re-create the edit using the video master tapes; if effects and titles are to be added, they too are integrated in online.

OpenDML: Open Digital Media ("Language"); a PC-based consortium of video professionals dedicated to improving Microsoft's Windows NT & 95 and Video-For-Windows to be more streamlined for high-end video product needs.

PAL: Phase Alternating Line. The European video broadcast standard consisting of 25fps with 625 horizontal lines per frame.

PCI: a pathway around PC computers (and eventually, other types) with reasonably high bandwidth. Peripheral Component Interface.

perf: a sprocket hole in film. Usually 4 per frame, but occasionally 3. Short for *perforation*.

picon: an expression for a "picture icon" *e.g.* as seen on systems to represent shots.

pixel: refers to the smallest unit of a reproduced image. For digital video, it is a sample of digital image information composed of *luminance* and *chrominance*. Short for *picture element*.

post-production: all work following the shoot of a production; begins when the film or video leaves the set and ends with the final release of the project.

pre-production: the work going into a film or video production, beginning with development of an idea and concluding with the shoot.

prosumer: a high-end consumer using equipment that borders on professional.

3:2 pulldown (also 2:3 pulldown)**:** the method for telecineing 24fps film to 30fps videotape. It involves the transfer of each film frame to alternatingly 2 video fields, then 3 video fields. A 3:2 pull down expands 4 film frames to 10 fields of video or 5 video frames.

QuickTime: a format for digitized moving video for Apple Macintosh computers. Like other special formats (e.g. TIFF, PICT, EPS), QuickTime "movies" can be used with many Macintosh applications and placed in many kinds of documents.

RA: Random Access, as in "RA editing systems." RA systems are usually but not always also nonlinear.

RAID: Redundant Array of Inexpensive Disks. A method of using a number of disks in parallel to increase the effective data transfer rate of a single disk. This means can be used to radically improve the image quality that can be managed with a digital video system. The "inexpensive" disks tend to be MO or hard disk drives.

RAM: Random Access Memory. Active but temporary computer memory. As opposed to ROM.

Rank: Slang for Rank Cintel, a major manufacturer for telecine machines.

reboot: to restart a computer. "Soft" reboot does not interrupt electricity to the computer and is considered safer. "Hard" reboot is turning the machine off and then on again.

render: to generate a video effect from its elements. Many video effects involve complex mathematical calculations to visualize and these can take even powerful computers a bit of time; consequently, effects are often visualized in low quality and at some point need to be fully generated for inclusing in a project. On digital systems, even simple dissolves often need to be rendered the means to visualize the effect are unavailable.

RGB: an abbreviation for red, green, and blue — the three additive primary colors used to construct video images.

ROM: Read Only Memory. Permanent computer instructions (data) on a chip.

SCSI: Small Computer System Interface. The output of a PC that generally connects to the hard disk, but it may connect other things. Pronounced "scuzzy."

SIMM: Single In-line Memory Module. A unit of RAM for PCs. Similar to DIMMs (Dual In-line Memory Modules) used in some Power Macs.

SMPTE: Society of Motion Picture and Television Engineers.

software: computer components with a no real physical form; software is a coded series of instructions that can be written out or recorded onto memory devices (chips, disks, etc) but is itself considered intangible, as opposed to hardware.

telecine: a film-to-tape transfer machine. Common manufacturers include Rank Cintel and Bosch. To telecine is to record film onto videotape. Usually pronounced "tela-sinee," but less commonly as "tela-seen."

turnkey: a term for describing equipment; it means that you could simply plug it in and turn it on, without needing any other equipment to make it work. The opposite of turnkey might be "batteries not included" or "some assembly required."

virtual editing: editing without using a master "record" videotape — as in digital systems.

VITC: vertical interval timecode — video picture encoded, and very accurate, timecode. Usually pronounced "VIT-see".

VTR: videotape recorder, as opposed to a video cassette recorder (VCR).

wavelet: a complex mathematical scheme for compression of video

YIQ: the American transformation of RGB color into its luminance ("Y") and chrominance ("I" and "Q") signals. Other transformations include: Y, Cr, Cb, (which is the same as Y, R-Y, B-Y) and HSB (hue, saturation, brightness).

YUV: a conversion of RGB color space, like "YIQ." "Y" is the luminance signal; "U" and "V" together make up the chrominance signal. YUV is usually used to describe PAL video signals.

QUICK CONTACTS

Trade Publications:
A/V Video: 701 Westchester Ave., White Plains, NY 10604 (914) 328-9157
Digital Video: 600 Townsend St. Suite 170 E, SF, CA 94103 (415) 522-2400
Film and Video: 8455 Beverly Blvd., Suite 508, LA, CA 90048 (213) 653-8053
New Media: 901 Mariner's Island Blvd., Suite 365, San Mateo, CA 94404 (415) 573-5170
Me!dea Magazine: 1033 Battery St., #206, San Francisco, CA 94111 (415) 989-9429
Millimeter: 16255 Ventura Blvd., Suite 300, Encino, CA 91436 (818) 990-9000
On Production and Post-production: 17337 Ventura Blvd., Suite 226, Encino, CA 91316
    (818) 907-6682
Post: 25 Willowdale Ave., Port Washington, NY 10050 (516) 767-2500
Videography: 2 Park Avenue, New York, NY 10016 (212) 779-1919
Video Systems: 9800 Metcalf Ave., Overland Park, KS 66212 (913) 341-1300

Training (also contact individual system manufacturers directly):
American Film Institute-Professional Training Division: 2021 N. Western Ave.,
    LA, CA 90027 (213) 856-7690 or (800) 999-4AFI
Full Sail Center for the Recording Arts: 3300 University Blvd., Winter Park, FL 32792
    (407) 679-6333 or (800) 221-2747
Maine International Film & Television Workshops: 2 Central St., Rockport, ME 04856
    (207) 236-8581
UCLA Extension: 10995 LaConte Ave. LA, CA 90024 (310) 825-9971
Weynand Training: 6800 Owensmouth Ave. #345, Canoga Park, CA 91303 (818) 992-4481

Post-Production Facilities (with extensive nonlinear or video-film experience):
Editel: 222 E. 44th St., NY, NY 10017 (212) 867-4600
Encore Video: 6344 Fountain Ave., Hollywood, CA 90028 (213) 466-7663
Image Transform:4142 Lankershim Blvd., N. Hollywood, CA 91602 (818) 985-7566
Laser-Pacific: 809 North Cahuenga Blvd., Hollywood, CA 90038 (213) 462-6266
Magno Sound and Video: 729 7th Ave., NY, NY 10019 (212) 302-2505
Modern VideoFilm: 4411 Olive Street., Burbank, CA 91505 (818) 840-1700
Post Group: 6335 Homewood Ave., Hollywood, CA 90038 (213) 462-2300
Tectis Group (Teletota): 2, rue du Bac, 92158 Suresnes Cedex, France (1) 40 99 50 50

Other Handy References:
Academy of Motion Picture Arts & Sciences, reference library: (310) 247-3020
American Cinema Editors (ACE): 1041 North Formosa Ave., West Hollywood, CA 90046
Christy's Post Production Ctr: 135 North Victory Blvd., Burbank, CA 91502 (818) 845-0008
MacUser magazine: 950 Tower Ln., 18th floor, Foster City, CA 94404 (415) 378-5600
MacWorld magazine: 501 Second St., San Francisco, CA 94107 (415) 243-0505
Motion Picture and Videotape Editors Guild (IATSE #771): 353 West 48th St., New York,
    NY 10036 (212) 581-0771
Motion Picture and Videotape Editors Guild (IATSE #776): 7715 Sunset Blvd., Suite 220,
    Hollywood, CA 90046 (213) 876-4770
SMPTE: 595 W. Hartsdale Ave., White Plains, NY 10607-1824 (914) 761-1100

I N D E X

Michael Rubin is an editor, consultant and teacher, currently residing in Santa Cruz. Graduated from Brown University with an Sc.B. in Neural Sciences, he joined Lucasfilm Ltd.'s new division, The Droid Works, where he was instrumental in the development of training for the EditDroid, and in working with the Hollywood market in evaluating and adopting new electronic systems for film.

photo by Jennifer Kuhn

mhr@cruzio.com

Upon The Droid Works' close in 1987, Rubin worked with the digital audio start-up, Sonic Solutions, which was developing new methods for Compact Disc pre-mastering and digital noise reduction/elimination techniques.

Late in 1987 Rubin joined CMX as a staff editor, working in the development and initial release of their CMX 6000 laserdisc-based editing system, and for a number of years supported its full-time use at The Post Group, in Los Angeles. He left CMX in 1989 to freelance.

He has been an instructor at UCLA Extension courses and seminars, a panelist at UCLA and in the Women in Film conference, a moderator for ITS panels, an advisor and instructor for the AFI-Apple Computer Center, and a teacher to hundreds of professional film editors.

Rubin has supported the use of electronic editing systems on many projects: commercials and music videos; the first feature film edited on the 6000, *Perfect Victims*; considerable film-originated television programming such as *Baby Boom* and *Dolphin Cove* and the mini-series *Lonesome Dove*. Rubin assisted Academy Award winning editor Gabriella Cristiani in her electronic post production of Bernardo Bertolucci's *The Sheltering Sky*, and was a principal editor on the Paul McCartney concert film *Get Back*, both on the CMX 6000. He was also an editor on one of the first television series to be cut on the Avid/1 media composer, *Love and Curses*.

Rubin has lectured internationally on nonlinear and electronic film editing: as a guest of the EFFECTS conference in Hamburg, Germany; at the International Television Symposium in Montreux, Switzerland, for the Directors Guild in Hollywood, California. Today, Rubin continues to teach, write and consult. Along with his work as a futurist, he and his wife Jennifer founded *Petroglyph Ceramic Lounge*™, a growing business in the SF Bay area. Along with everything about his Macintosh, he enjoys writing, painting, photography, films and travel.

O R D E R   F O R M

TO: Triad Publishing Company
      PO Box #13355
      Gainesville, FL 32604

Please send me _____ copies of *NONLINEAR 3*, at $29.95 per copy, plus shipping and handling of $7.00 1st book, $1.00 each add'l book (add 6% sales tax for Florida orders). Credit card orders (Visa or MC) include cardholders name, card number, and exp. date.

Name: _____

Address: _____

City/State/Zip: _____

Number of books _____   Total amount included $_____

Foreign orders: Payable in U.S. funds drawn on a U.S. bank (write for shipping charges, and indicate air or surface) or by credit card (Visa or MasterCard). Include cardholder's name, card number, and exp. date. Visa and MasterCard orders may be faxed TOLL FREE to 1-800-854-4947 All orders shipped promptly.

Prices subject to change without notice.